Cost-effective maintenance of railway track

Proceedings of the conference *Cost-effective maintenance of railway track* organized by the Institution of Civil Engineers and held in London on 25–26 June 1992

 Thomas Telford, London

Conference organized by the Institution of Civil Engineers

Organizing Committee; R. A. Vickers (Chairman), Mott MacDonald, S. Kercher, M. J. Gobey, London Underground Ltd, J. H. G. Cook, Transmark, A. Wray, Overseas Development Administration, J. Buekett, Consultant, and J. Young, formerly British Steel Track Products

A CIP catalogue record for this publication is available from the British library

ISBN 0 7277 1930 0

First published 1992

Published on behalf of the organizers by Thomas Telford Services Ltd, Thomas Telford House, 1 Heron Quay, London E14 4JD.

Printed in Great Britain by Redwood Press, Melksham, Wilts.

Contents

Opening address. L. Thompson   1

Discussion on opening address   9

Specifications

1. A review of knowledge on vehicle – track interaction.
   B. EICKHOFF   11

2. Standards for track components. G. G. LEEVES   27

Discussion on Papers 1 and 2   37

Track condition

3. Maintenance tolerances – desirable and minimum values.
   P. FORBES   43

4. Upgradation of track for higher speeds and rectification of long
   wavelength defects in track geometry. M. SESHAGIRI RAO   51

5. Inspection, monitoring and measurement of track condition.
   D. BALLINGER   63

Discussion on Papers 3 – 5   81

Hazards and structures

6. Bridges and structures. M. ARSHAD   89

7. A review of the effects of natural damage. D. SPERRING   105

Discussion on Papers 6 and 7   115

Maintenance equipment

8. Manual methods - the alternatives of length gangs and mobile
   gangs. J. K. MUSUVA   117

9. Maintenance equipment — users. J. SVENDSEN   129

10. Maintenance equipment — suppliers. A. LEYLAND   133

Discussion on Papers 8 – 10   **147**

Organization

11. Planning of the work. A. J. LINDSAY   **149**
12. Track rehabilitation in developing countries - direct labour or
    contractor? R. T. MASTERS   **159**
Discussion on Papers 11 and 12   **175**

Case studies

13. Cost-effective maintenance of railway track — Indian
    Railways. M. RAVINDRA   **181**
14. Cost effective maintenance of railway track — Scotrail.
    M. CHORLEY   **191**
15. Cost-effective track maintenance on Queensland Railways.
    F, BELL   **203**
Discussion on Papers 13-15   **219**

Opening address

L. S. THOMPSON, Railway Adviser, The World Bank

I very much appreciate the opportunity to address you. As I am sure we are all only too well aware, railways are coming under increasing pressure to produce more for less, and do it better, in the face of constantly growing competition from other modes. This is true in the EC, where the coming of Europe 1993 and the recent European Commission order on railway organization and competition will combine to bring about a real revolution in freight competition in an expanding Europe. It is also true in the economies in transition in Central and Eastern Europe where the railway's traffic has traditionally been enshrined within the 'plan' and not really available to competitors. It holds equally for all of the developing world where governments are looking to a market-based economy for economic growth — and the foundation of a market economy is stiff, unrelenting competition in all sectors. While it is customary in the USA, at least, to begin speeches with a point of levity, I am not going to do that here: I want to underscore the seriousness of the challenge and the need for an effective and timely response from the railway community.

There is at least some good news: the railway cause is not hopeless. The tools and technology for winning this battle are available. The bad news is that stability will not be easy to achieve, and it will not be won by looking to the past. Change is in the air, and most of you will be in the thick of it, which may be either good or bad news, depending on your willingness to respond.

The World Bank believes that railways are important; they should, they must, be a part of the fulfilment of the transport requirements that economic growth and restructuring will bring. Countries allowing inefficient or ineffective rail service do so at a very high cost to the growth of the economy to the nation's fiscal health, and to the national competitive position in a world where international trade is more and more important. The stakes in having effective railways are not low. We are now lending money to railways in every region of the world, and I expect we will continue to do so. Where there is an opportunity for the money to be put to a productive purpose, the Bank is ready to become involved, and to persist in its role until progress is made.

I hope this talk can set something of a backdrop for the conference — that is, establish some sort of perspective against which the real importance of some of these issues can be appreciated. I hope it will surprise no one in this audience (although I know some who are always shocked to hear it) when I

1

stress that railways do not exist because we like to have them or because they employ us. All of us find comfort in the sound of the whistle in the night, and of course there is something uniquely interesting about railway work, history and culture. This said, the World Bank supports railways when (and if) there is an opportunity for the railway to provide a necessary service to the economy which cannot be obtained more effectively any other way. Bluntly, railways survive when they are better than the alternatives. Better in this case means offering the service at higher quality or lower cost, or both, which meets the customers' needs: better does not mean empty regional passenger trains, or slow, unreliable freight services. The Bank prefers to concentrate its railway lending in cases where the role of the railway has been agreed and is clear.

By the same token, none of us would build or maintain railway track because it pleases us to know it may someday have a use. We build track because there is traffic for it to carry, traffic which is actual, stable and predictable for the future. And we (should) maintain that track at the absolute minimum cost needed to carry the traffic. Track without traffic is a waste of capital and every excess dollar sunk in track is a dollar not available to buy new wagons or locomotive spare parts. Track which is maintained to a level, or in a way better than the traffic requires is an invitation to more efficient competitors, with lower costs, to take the traffic away.

I cannot stress too strongly, therefore, that cost-effective track maintenance has two facets. At the most basic level, of course, we need to put and keep the track in the desired condition in the least costly way. Engineers unfortunately tend to want to concentrate on this point because they are comfortable with it. But 'desired' does not necessarily mean perfect. In the case of the TGV, it might (or nearly so): for a branch line in rural Zambia, it means the absolute minimum effort needed to keep the line in place and carry one light train per week. Business purpose must come first in defining appropriate track quality: engineering efficiency comes next in meeting business purpose. However elementary this may appear on paper, it is a common point of contention in our railway lending.

In order to put today's railway discussion into better comparative perspective, I would like to refer to some excerpts from the Bank's railway database. The Bank's database contains many of the basic facts about a railway, including track, equipment, operating results, financial result, and basic productivity measures. It has at least some information in a common format concerning 82 railways in 71 countries. While the database is not error free, it can reasonably be used to get a good, overall picture of the capital plant and operations in the world's railways.

Overall, the worlds railways produced roughly 7.3 trillion tonne kilometres of freight service and 1.7 trillion passenger-kilometres (pkm) of passenger service in 1988. The railway of the former Soviet Union alone produced about half of all of the world's TKM and 24 % of the PKM. Next in line, the US class I railways generate about 19 % of the world's TKM (but only 0.5 % of the world's PKM). By contrast all of the developing railways together account for only 23 % of the world's PKM but do produce 49 % of the PKM (and China and India which are the world's third and fourth largest railways in traffic produced, account for roughly 70 % of the developing railways' TKM and PKM.

The statistics underline the wide range in types and intensity of activity among the world's railways. If, for example, ongoing efforts for track maintenance and renewal are largely determined by plant size and traffic activity, then China Railways with 53,000 km of line, and an average traffic density of 24.9 million Traffic units (TU, which is the sum of TKM and PKM) per year per kilometre of line (the second highest in the world after the USSR) concentrated largely in freight traffic, is a massive daily operating challenge comparable in difficulty with any railway in the world. On the other hand, many developing railways operate with a traffic density of far less than one million TU/yr/km, which poses quite a different question of operating economy. China Railways, like many of the Eastern European railways, is capacity focused, with throughput being a real concern. Many other railways will focus on carrying the traffic offered for minimum cost. Much the same can be said of employment.

Another managerial and engineering challenge I should mention is found in locomotive fleet and availability. Locomotive availabilities range from slightly over 90 % under the best of circumstances to less than 50 % where problems are especially acute. Like body temperature in human health, I have come to believe that locomotive availability rates are probably the most sensitive single index of the operating health of the railway. As with elevated thermometer reading, there can be many reasons for depressed locomotive availability, including a shortage of spare parts or of financial resources, poor maintenance facilities, skills shortages, a technically inadequate locomotive design for the service required, or perhaps just a simple inability to take superannuated locomotives off the roster. Whatever the reasons, low locomotive availability signals a need for remedial action. For this conference, suffice it to say that track has little use without locomotives to haul the traffic — and there are a number of developing railways where the chief constraint on traffic is locomotive power.

A different issue emerges from a look at what is probably the most frequently discussed (if not necessarily most significant) comparative performance measure, labour productivity, measured by annual TU per employee. This ranges from well below 100,000 in a few countries to nearly 9 million for the extreme case of the freight oriented, high labour cost railways in the US. There are, again, sufficient reasons for variation in this index (low wages, for example, would often permit lower TU/employee) to require judgement in its interpretation. One variable in particular, appears to have an impact on output per employee, and that is the balance between passenger service and freight service. There is a very wide difference in the role played by developing railways (as there is in the railways in developed countries) in this respect. Aside from its impact on output per employee, this difference has clear implications for the types of rolling stock required, as well as for the levels of maintenance needed for both equipment and track.

A final, critical issue is traffic trends. When traffic is rising (short of real capacity constraints), many problems can be resolved simply through growing out of them if resources are available. Managing against a shrinking traffic base is an entirely different question because costs rarely fall as quickly as (and employee morale often falls more quickly than) traffic. Not only are total traffic levels changing on many railways but, in many cases, TKM and PKM are growing (or shrinking) at different rates, adding the additional management issue of a change of role as well as of size.

The Bank believes that railways can play an important role in the solution (and unfortunately also the creation) of significant economic problems. Whether the railway contribution is positive or negative depends on how well the role, goals and efficiency of the railway match the needs of the economy — that is, does it have an appropriate business purpose. An effective definition of purpose requires two steps, a good railway strategic plan, and a workable relationship between railway and government, often expressed in a contract plan or performance agreement.

The strategic plan is the primary document for clarifying the desired role to be played by the railway, as opposed to other modes. It must be based on an agreed set of broad economic scenarios for the country, and a well developed definition of the alternative objectives for the railway, including its expected role in carrying both passengers and freight. For the preferred role or roles, a consistent set of policies and regulation is needed for both railway and government to follow in fulfilling this role.

In the strategic plan, the railway should approach its activities commercially: basically, if customer revenues will not pay for a service (assuming the railway operates efficiently) then the railway should not be asked to provide the service from its own resources. It is up to government to define any social or environmental benefit from rail service as compared with other modes and, if the cost is acceptable, the government can choose to support the railway. In this formulation, the railway should view the government as customer, much like any other. The strategic plan is also the place to raise and resolve a host of other issues, such as imposed surplus employment or unfair subsidies to other modes, which act to impede the repositioning of the railway.

The performance agreement should then clearly define the authority, responsibility and performance targets of railway and government in implementing the strategic plans goals. The railway's responsibility to be efficient, to provide quality service, to compete effectively to stay within financial targets, to deal fairly with government as customer, all must be set out and agreed, as must be the authority of the railway to act without interference within its sphere of decision-making. The Government's role as definer of social role, fair regulator, provider of operating payment, supplier of agreed capital and partner in transitional issues, must also be set forth in a way which generates from commitment at the political level. Both sides of the agreement must be assessed: plans without authority and resources cannot be implemented; authority without plans has no direction; government intervention without responsibility destroys management accountability.

Very few economies lack a useful potential role for rail transport, although the proper future role (indeed, the appropriate present role) may be quite different from the railway's actual services. Modem developed economies facing growing challenges of congestion and pollution, can increasingly make use of the capacity and cleanliness of rail both for freight and passengers which in Western Europe, includes high speed rail. Developing economies often depend on the longer distance movement of bulk commodities and the low-cost transport of passengers, roles for which rail can be naturally fitted.

Once the business and social (if any) purpose for the railway is defined and agreed, and a workable relationship among railway, government and market has been reached, we can usefully talk about track. Although the variables in track construction and maintenance are far too complex to permit summarisation, I believe that the two most important determining questions for tracktype, capacity and maintenance quality are the balance among the types of traffic being carried (freight, intercity passenger, suburban passenger, and high speed passenger, among many others) and the intensity of traffic.

The first major determinant, balance among types of traffic, has many facets, some of which I have already discussed. Freight service typically uses slow, heavy trains which stop infrequently, accelerate slowly and have quite heavy axle loads. The usual freight customer wants lower cost but reliable service. Under these conditions, the track needs to be sturdy, but does not have to be built or maintained to extremely high geometric standards. By comparison, intercity passenger trains are faster, lighter, stop less frequently, accelerate more quickly, and have lighter axle loads. Unlike freight, intercity passengers are also sensitive to track quality as well as strength. To complete the list, suburban (or local, rural) trains tend to be short, light, multiple-stop, fast and have high acceleration. Finally, high speed passenger trains exaggerate the situation: they need to be effectively straight (minimum curve radius of 4 km) and extremely high quality.

Overall, the higher the share, speed and frequency of passenger service, the better the track must be, and the more expensive the maintenance required. At high shares of freight traffic, more importance can be attached to purely economic considerations, and quite wide variations in track quality are permissable. As all of you know, the worst of all worlds comes when we have to do all at once, on the same track. Combined operations produce both overbuilt and over maintained track.

A more subtle effect of the freight versus passenger balance is the effect it has on the relationship between government and railway. With few exceptions, intercity passenger services are consumers of public funds (even fewer exceptions for suburban rail services), which means that passenger railways are guaranteed a high level of government interest and, unfortunately a high risk of government interference. Construction and maintenance of passenger railways must be conducted with a different degree of accountability and reportability than for predominantly freight. As well, the performance agreement must reflect the higher level of government interest in safety and reliability of passenger service than for freight.

The world's railways have considerable variation in their degree of passenger orientation. In North America, rail passenger service has a small share (less than 1 %). Other countries, Russia and Mexico for example, are also predominantly freight carriers. Most Western European and CEE railways have shares of passenger service in the 30 to 60 % range. Asian railways often run the highest passenger shares, with some (Japan, Bangladesh and Sri Lanka) running over 90 % passenger operations. Japan, France, Germany, the UK, the US and Spain are the only countries with trains faster than the 200 km/h mark usually acknowledged to be high speed, but only the US, UK and Germany mix high speed and other traffic.

The second factor, density or intensity of traffic, significantly affects construction designs and maintenance effort. Heavy use requires increasing strength, mechanisation and precision, while light usage focuses attention on efficient use of traditional methods where extensive new investment is not justified.

There is wide variation in the traffic density of different railways. The highest two, USSR and China, operate at traffic densities (29.4 and 24.9 million net TKM/Km of line respectively) which are roughly ten times the average West European practice, and a much larger multiple of typical developing railway experience. At these high densities, time slots for track maintenance are few and far between, and track access must be carefully planned. At these higher densities also, the cost of inadequate maintenance (accidents, slow orders, rough ride) will be felt sooner, and will rapidly progress to serious levels. At the other end of the scale, every maintenance dollar can, and must, be held to a minimum because the economies of traffic density are not available to compensate for maintenance cost.

Track carrying heavy traffic needs high quality renovation frequently. Experience is showing that operating economies can be achieved on heavily used track, even at extremely high axle loading (the AAR test in Pueblo have assessed gross wagon weights of as much as 141 metric tonnes on 4 axles) but the track must be strong (60 kg/m or more), and the condition of the track must be maintained at a high level. This means high quality rail (in many cases, hardened or alloy rail), sleepers in good condition (concrete serves well) and elastic fasteners, and excellent ballast and other track materials — maintained to strict standards. Although lightly used track does not necessarily justify heavy maintenance or complex control systems, it can often be usefully and economically upgraded using re-welded, cascaded rail with ballast cleaning and spot replacement of sleepers.

Track with high traffic density is also very expensive to take out of service because of the disruption this causes to traffic, and depends on mechanised methods for effective, minimum down time maintenance, including high volume line and switch tampers, mechanised track replacement systems, and unitary switch replacement methods. Fortunately, the equipment needed to minimise out of service time is also the right equipment for high quality, high strength work. Lightly used track can generally be more effectively maintained with existing methods, but with more extensive use of modern maintenance planning and programming methods. In addition, the track used in carrying passengers, especially at high speeds, must be seen as a precision instrument receiving careful attention at all times, essentially without regard to the traffic levels, as a result of ride quality and passenger safety considerations.

Whatever the usage, certain other themes are becoming increasingly common in effective rail management. Concentration of traffic is important, because concentration produces the opportunities for economies of scale and design. In all cases, appropriate information collection and reporting is vital: we can only manage what we can measure and control. In larger railways this implies heavy use of computers and communications: in smaller railways this requires adroit and disciplined use of information already available.

Another theme of growing importance in track construction and maintenance (among other areas) is enhancing the role of the private sector in railway services and functions. Most railways are cautious about looking out of house for help for various reasons, including the risk of having outsiders on the track and a traditional policy of seeking autarchy because railway skills are so specialised and proper maintenance is so important: another reason, truth be told, has been the resistance of railway labour to the introduction of outside workers in what has often been a comfortable and highly protected employment market.

There is now a fully established body of experience which shows that, under the right circumstances, track construction and maintenance can be performed more efficiently under contract with the private sector than in-house. This can particularly be true when, as is increasingly common, the maintenance effort requires expensive and specialised equipment and techniques requiring skills and diversity of application which are beyond what any individual railway can provide. For some functions, such as rail grinding, this has long been common: it is becoming more common for others, such as track laying ballast cleaning. There is little doubt in my mind that increased use of contracting for track maintenance will continue.

The real bottom line though, is the fact that highway and air competitors for capital budgets are becoming increasingly capable of effectively presenting and supporting their needs. In this struggle for investment — and for existence — railways will have to redouble their efforts to ensure that they use what they have well before they ask for, or receive, more. A key factor in this cost equation will be track construction and maintenance. For railways there is both more opportunity for success, and less margin for error, in the future, and I think cost-effective track maintenance will be one of the determinants of success.

Discussion on opening address

F. I. MAU, Vice-President Operations, BHP Rail Products (Canada)Ltd
In identifying which project to support, does the World Bank consider first cost or life cycle cost? If the latter is true, does the World Bank consider annualized cost of an asset having infinite life, or the present cost of an asset for a specific life? What life would then be considered, and what rate of return would the bank expect to achieve for the project to be attractive?

L. S. THOMPSON, Author
The World Bank attempts to do economic and financial rate of return (ERR and IRR) calculations which, if done correctly, would inherently include life cycle costing. IRR/ERR calculations attempt to use the appropriate life for the item in question although, as is well know to practitioners in the field, the lifetime question is difficult. There is no single satisfactory rate of return: in some economies where capital is extremely scarce and the level of development low, rates of return above 30% are typical and form a sort of cutoff for the economy; in more developed economies, rates of return as low as 10% could be acceptable.

1. A review of knowledge on vehicle – track interaction

B. M. EICKHOFF, Vehicle Dynamics Unit, British Rail Research

Introduction

The running of a railway vehicle over a length of track inevitably produces dynamic forces both on the vehicle and on the track. The magnitude and form of the forces depends on the vehicle design and maintenance and on the track design and geometric quality. Neither the vehicle nor the track can operate in isolation, the interaction between these two components of the railway system is of fundamental importance.

The vehicles affect the track in that the forces produce noise and vibration, geometric deterioration, component damage and wear. The track affects the vehicles in terms of safety risk and vehicle ride as well as component damage, deterioration and wear. An understanding of this interaction is of considerable importance both to civil and to mechanical engineers in ensuring that their part of the system is suitable for its job. The principles involved are independent of the type of railway system, but the relative importance of different effects clearly depends on the speed of operation, quality of track and range of curvature.

A considerable amount of work, both experimental and theoretical, has led to a good understanding of the main areas of vehicle–track interaction and to the development of usable prediction methods. This Paper looks at some of the work that has been carried out, and the knowledge that has been derived, concentrating on the effect on the track rather than on the vehicles. The aim of the document is to highlight some of the mechanisms and questions rather than to provide specific answers, which can be given only on a case by case basis.

Vertical forces

Static effects

The quasi-static vertical loading of a length of track as a vehicle passes is the simplest form of vehicle–track interaction. The load from each axle is distributed over a few sleepers, depending on the details of the track structure, and the track and its foundation must be capable of supporting the train weight. This situation is not, however, quite as trivial as it at first appears.

The load on each wheel has to be transferred to the rail through the Hertzian contact patch, typically an area of about 50 - 150 mm^2, and the stresses generated in this comparatively small contact patch can be large and have to be limited to avoid rail failure. The commonly used ratio of axle load to wheel diameter is one way of limiting these stresses. A more comprehensive measure would take into account a realistic range of wheel and rail cross-sectional shapes, and hence contact area shapes and sizes, and possibly make allowance for the vehicle suspension type.

Straight track forces

No length of track is perfectly smooth; there are always irregularities present, within the rails themselves, in the track support stiffness, in the ballast distribution etc.. Under the passage of vehicles these irregularities give rise to dynamic forces and these forces in turn lead to deterioration of the track structure and geometry.

General track irregularities occur at a range of wavelengths and can cause dynamic behaviour in the vehicle which in turn causes magnified track forces. A vehicle passing at a given speed will see a dominant wavelength as a particular frequency input and knowledge of the wavelengths present within track irregularities is important to avoid potential problems.

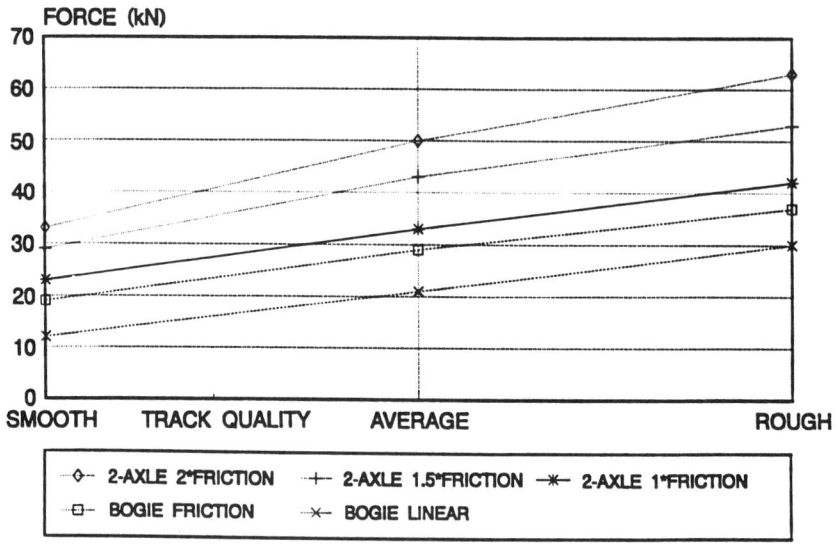

Fig. 1. Freight vehicle vertical track forces

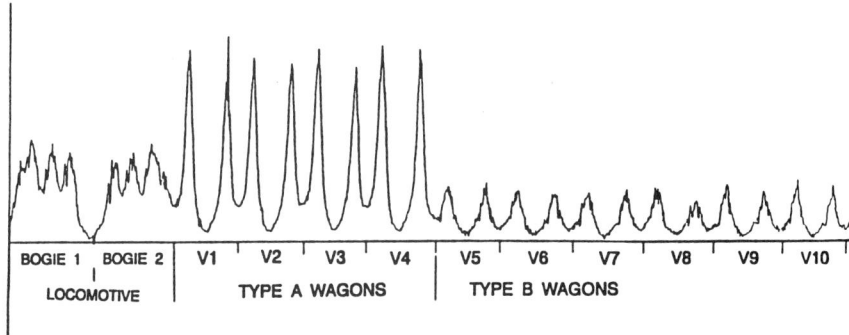

Fig. 2. Measured sleeper strains

Any railway vehicle has a number of natural modes of vibration, forms of motion in which little force is required to produce movement. If the dominant wavelength in the track combined with the operating speed of the vehicle happens to coincide with the frequency of such a mode of the vehicle then very large movements can be generated which lead to passenger discomfort and in extreme cases to derailment.

The suspension design of the vehicle can have a significant effect on the magnitude of the dynamic forces that are produced when passing over track of a given roughness. Fig. 1 shows the results from calculations of the dynamic vertical forces from a number of different freight vehicles on track with a range of roughnesses representative of BR freight track. The level of suspension friction within the two-axle vehicle makes a considerable difference and the bogie vehicle with the linear, viscous damped, suspension can be seen to produce much lower forces, particularly on the smoother track.

It has also been observed that vehicle suspension has a significant influence on track vibration and on the amount of noise and ground-borne vibration generated. Fig. 2 shows strains measured in concrete sleepers under the passage of a locomotive and a train of two types of freight wagon of the same nominal axle load. The difference in strain level is extremely marked.

Rail irregularities exist at a range of wavelengths and one of the most well known is the short wavelength corrugation, typically in the range 50 - 100 mm. These are a serious problem in causing high frequency sleeper vibration leading to loose fastenings and deterioration of sleepers and ballast as well as to noise generation. A number of explanations for this phenomenon have been developed but, so far, no fundamental solution has been derived and rail grinding remains the only way of removing corrugations.

Wheelflats and rail dips

Two particular types of irregularity that can give rise to large dynamic vertical forces are damaged wheels (wheelflats or out-of-round wheels) and rail dips (either at joints or welds). In both cases a relief of load takes place at the irregularity and, if it is severe enough for loss of contact to occur, this will be followed by an impact.

The vertical forces generated at dipped joints are well known, often resulting in a persistent maintenance problem at particular locations. These forces have been characterised as 'P1' and 'P2' forces with the P1 being the vibration of the wheelset mass on the Hertzian contact spring and the P2 being vibration of the wheelset and rail mass on the track support stiffness. In order to limit these forces restrictions are usually placed on the vehicle "unsprung mass", that part of the mass which is below the primary suspension.

A significant difference between a wheelflat and a dip is in the length of the irregularity. A severe wheelflat would be only 100 - 150 mm long whilst a dipped weld would initially be perhaps 600 mm long, lengthening to several metres after the passage of numerous trains. This leads to a difference in the force response as loss of contact is much more likely for a wheelflat than for a dip. Calculations suggest that only a relatively small wheelflat, around the limit allowed on BR for some freight vehicles, can cause values of wheel–rail force, rail pad force and sleeper strain to be double those from a smooth wheel, whilst a severe rail dip is required to produce the same magnitude of effect.

It is also important to remember that the forces from dips are limited to the location of the original weld or joint whilst wheelflat forces occur every few metres throughout the journey, potentially damaging miles of track.

Wheel–rail force measurements of vehicles in normal service on BR have identified a number of wheel defects causing responses of 3 or more times the static values. This work has also shown that irregular, out-of-round, wheels can be at least as damaging as wheels with conventional flats, and are more difficult to detect without sophisticated measuring equipment.

Lateral forces
Wheelset kinematic behaviour

In order to consider the ways in which lateral forces are generated some understanding is needed of basic wheelset behaviour. The simplest railway

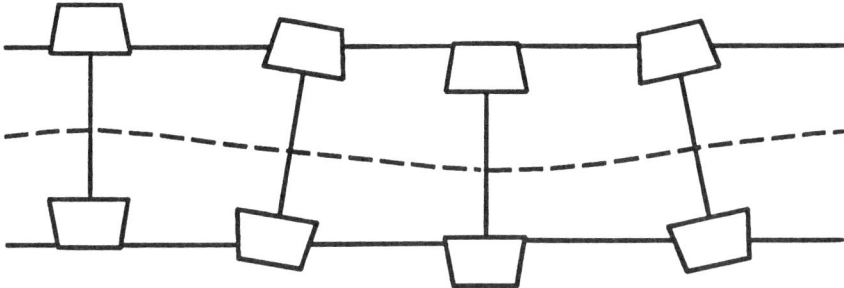

Fig. 3. Wheelset hunting behaviour

wheelset is made up of two coned wheels fixed to a rigid axle. On HR the standard cone angle is 1 in 20; typical values used on other systems are 1 in 40 or 1 in 30. This coned wheelset has the important property that if it is displaced from the track centre-line then the point of contact on each wheel moves, leading to a situation which is no longer symmetrical but where the local radius of one wheel is larger than that of the other. On straight track this has the effect of turning the wheelset back towards the centre of the track, leading to an oscillatory path (Fig. 3), and on curved track can help to "steer" the wheelset round the curve.

This basic hunting behaviour is dynamically unstable for a free wheelset and the magnitude of the oscillation would grow until limited by flange contact. Within a vehicle the instability is controlled by suspension springs and dampers until a 'critical speed' is reached, above which hunting occurs. Ideally this should be above normal operating speeds but in practice hunting does occur in some circumstances.

In practice most railway wheels are not simple cones, either because of wear or because of the use of a 'worn' profile. For a particular combination of wheel and rail cross-sectional profiles an equivalent cone angle or 'conicity' can be determined. As the wheelset moves within the flangeway clearance, the point of contact between the wheel and rail can be calculated and hence the local effective radii of the wheels and the conicity. The rail profile and the track gauge can be as important as the wheel profile in ensuring that the conicity is kept within design limits. If the railhead profile is too flat or the track gauge too tight then very high conicities can be generated.

15

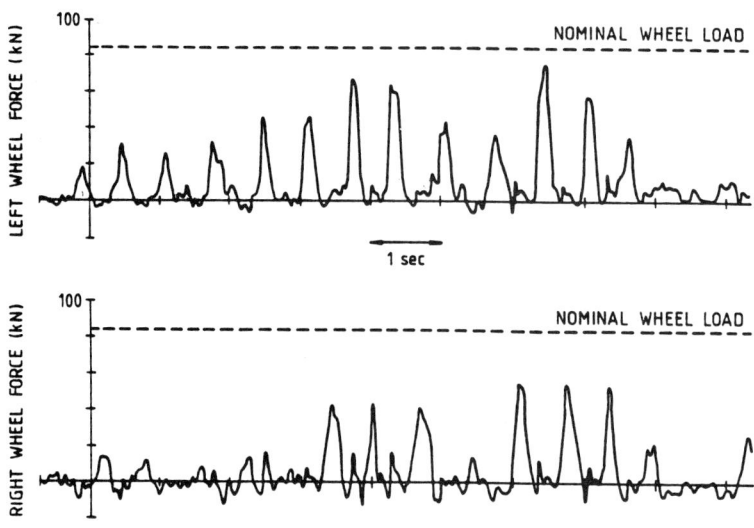

Fig. 4. Tippler wagon lateral wheel–rail forces

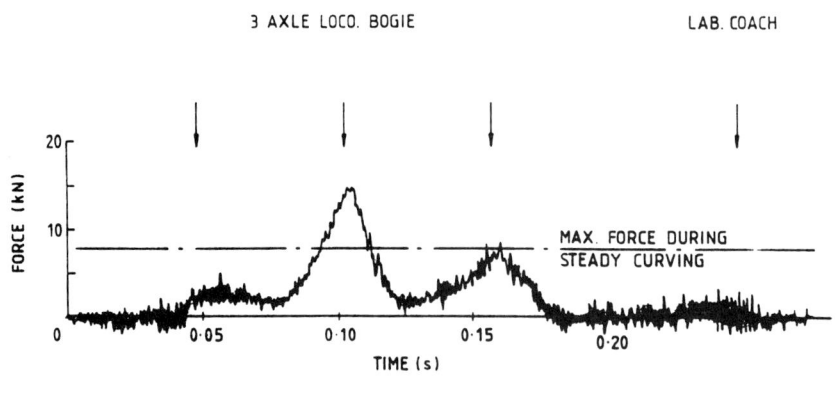

Fig. 5. Hunting of a three-axle locomotive bogie

Straight track

Very large lateral forces can be generated by vehicles hunting on straight track. Fig. 4 shows measured lateral forces from a two-axle freight vehicle. The cyclic behaviour can be clearly seen and the peak lateral force is approaching the nominal vertical wheel load. This hunting behaviour is not in itself a derailment risk as the large lateral forces occur at the same time as large vertical forces on the same wheel, but the track forces are severe. The presence of large track irregularities, either vertical or lateral, at a critical point in the hunting cycle, could lead to derailment.

Another example of lateral forces from a hunting vehicle on straight track is shown in Fig. 5. This shows lateral forces measured on instrumented fastenings on concrete sleepers during the passage of a three-axle locomotive bogie. Again significant forces are generated.

On straight track the wheelset's flangeway clearance of around +5 mm allows small alignment irregularities to be accommodated. Some lateral force will be generated but this is usually small unless large irregularities are present. On curves, irregularities are more significant as at least some of the wheelsets are likely to be in or near flange contact with the rail and therefore sensitive to quite small irregularities.

Switches and crossings

Each switch, or turnout, represents a large lateral irregularity for the vehicle to negotiate. In the turnout direction the train is required to change direction abruptly without the benefit of a transition in either cant or curvature and an angle (α) is required between stock rail and switch rail. On BR the effect of vehicle speed (V) is taken into account by using shallower angles for higher speeds such that V remains constant. This allows for inertial effects but fails to allow for the distance required for the wheelset to steer itself to the new direction. The result is that wheel–rail impacts are inevitable even on higher speed turnouts, with the magnitudes dependent on the track layout and the lateral unsprung mass of the vehicle. Locomotives with heavy traction motors are normally the worst type of vehicle as the motor is unsprung laterally.

Figure 6 shows measured and predicted lateral forces for a two-axle vehicle traversing a turnout. An impact occurs at the left wheel–rail contact followed by oscillation on the lateral track stiffness and large forces are generated. These forces must be allowed for in track design.

Curving

Steady state forces - balancing speed

When a vehicle is required to traverse a curve, lateral and longitudinal forces are generated at the wheel rail contacts. Fig. 7 shows the wheelset positions and the forces for a bogie vehicle on a comparatively sharp curve. Although the conicity effect discussed earlier will try to help the wheelsets to 'steer' themselves around the curve, the comparatively stiff vehicle suspension holds the wheelsets approximately perpendicular to the vehicle body. This leads to an "angle of attack" at the leading wheelset as it attempts to continue straight on and is forced round the curve by flange contact.

At the inside wheel of the leading wheelset this angle of attack gives rise to relative movement (or creepage) within the wheel–rail contact and this in turn gives rise to a lateral force. The force on the wheel is shown on Fig. 7. This force is pushing the wheelset towards the outside of the curve and has

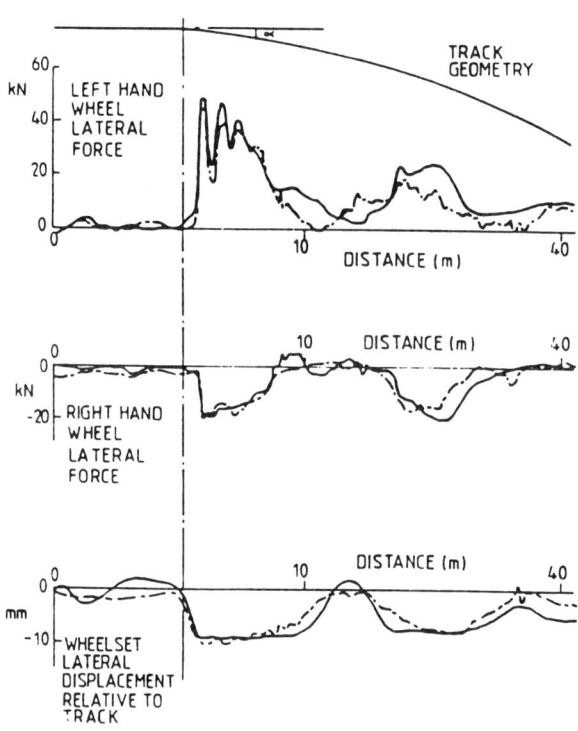

Fig. 6. Experimental and predicted results for a two-axle vehicle on a turnout

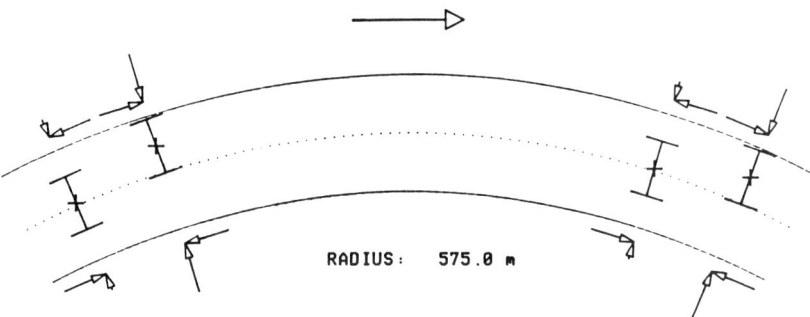

Fig. 7. Curving behaviour – stiff primary suspension

to be balanced by a flange force on the outer wheel of the axle (also shown on Fig. 7). These two forces together form the 'gauge spreading' force familiar to most railway civil engineers. In extreme cases these forces can actually force the rails apart, or damage track fastenings, to such an extent that derailment occurs.

On shallower curves, or with more flexible vehicle suspensions, the wheelsets can "steer" more successfully. Fig. 8 shows the same curve radius as Fig. 7 with a softer primary plan view suspension. The axles are aligned approximately with the curve and the forces are considerably reduced. Unfortunately, the use of a more flexible vehicle suspensions has a detrimental effect on the maximum speed of the vehicle and any design is therefore a

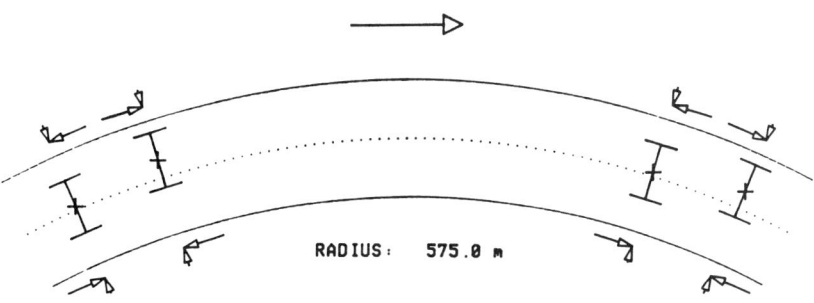

Fig. 8. Curving behaviour – soft primary suspension

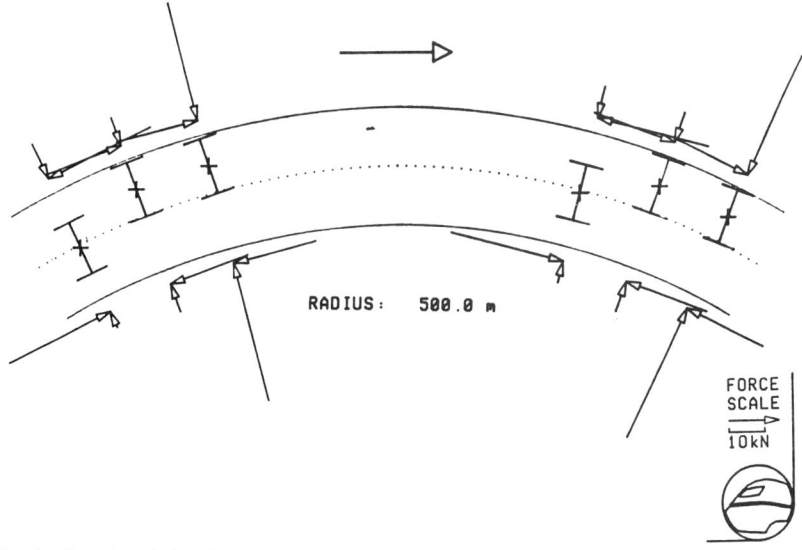

Fig. 9. Curving behaviour – six-axle locomotive

compromise between curve negotiation without excessive forces and the top speed required of the vehicle.

In general, more modern vehicles are designed to curve adequately down to curve radii of 1000 to 600 metres but few vehicles curve well on sharper curves unless special steering mechanisms are used and track design must take this into account.

A particularly severe set of curving forces can be generated by vehicles with three-axle bogies on sharp curves. The situation for a three-axle bogie locomotive is shown in Fig. 9. The centre axle, as well as the leading axle, has an angle of attack to the outer rail but, because of the restricted lateral clearance of the centre axle, the wheelset is unable to reach flange contact with the outer rail and the force is instead transferred through the bogie frame and reacted at the outer wheel of the leading axle. This leads to very severe forces at this contact which cause large amounts of wear and in extreme circumstances can destroy the rail–sleeper fastening system.

Track cant and cant deficiency

The description of curving behaviour given earlier dealt with curving at balancing speed, that is when the amount of crosslevel, or cant, on the curve exactly compensates for the lateral acceleration required to traverse the curve. Where this is not the case the vehicle has a net lateral acceleration which has to be reacted through the wheel–rail contact. In conditions of cant deficiency (higher speeds than balancing) this is towards the outside of the curve, in conditions of cant excess (lower speeds than balancing) it is towards the inside of the curve.

Another effect of curving at other than balancing speed is the load transfer from one wheel to the other with the inner wheel being loaded more in conditions of cant excess and the outer wheel in conditions of cant deficiency. The changes in curving forces that result from these vertical load transfers and superimposed lateral forces are complex but some general principles are worth mentioning.

The force on the low rail at the leading axle is a creepage, or friction, force and hence its maximum value is limited by the prevailing wheel–rail coefficient of friction and the vertical load on that wheel. Hence the load transfer from cant excess (slow speed running) can actually increase the lateral force on the low rail, and hence the high rail force that reacts it. This means that, contrary to normal expectation, the largest forces can sometimes actually be generated at the slower speeds. Table 1 shows the maximum lateral forces at the inner and outer wheels of the leading axle of a vehicle on a sharp curve for conditions of 4 degree cant deficiency, balancing speed and 4 degree cant excess for a number of different values of wheel–rail friction coefficient.

The lateral forces are expressed as a percentage of the static wheel load and it can be seen that the largest lateral force is 60% of the static wheel load and is generated in conditions of cant excess with a high friction coefficient.

Table 1. Maximum lateral forces on an axle

Wheel-rail friction (μ)	cant def (+) / cant excess (-) (degrees)	lateral force (% wheel load)		outer rail Y/Q
		inner rail	outer rail	
0.5	+ 4	39	53	0.44
	0	50	50	0.5
	- 4	60	46	0.58
0.4	+ 4	32	46	0.38
	0	40	40	0.4
	- 4	48	34	0.43
0.2	+ 4	16	30	0.25
	0	20	20	0.2
	- 4	24	10	0.13

At the higher friction levels the gauge spreading force is largest for the slow speed, cant excess situation. The values of derailment quotient (Y/Q) at the outer rail are also higher in conditions of cant excess, for values of corresponding to dry rails.

It can be seen from these results that the amount of installed cant on a curve has an influence on the behaviour of vehicles negotiating the curve and on the forces generated, not always in the ways that might be expected.

Wear and lubrication

The wear of wheels and rails on curves is a problem known to all railway engineers. Earlier discussion has indicated some of the mechanisms by which the creepages and forces are generated that give rise to wear. As these forces increase on sharper curves the wear clearly also increases but this effect is magnified by a change in the mechanism of the wear itself.

For low levels of creepage and force the mechanism is one of generation and breakdown of thin surface oxide films. Other than in locations of high traffic density this mild wear is seldom a problem. At higher levels of force and creepage the protective oxide films break down giving metal-to-metal contact and severe wear. This is characterised by the production of metallic wear particles, as seen in situations of hard flange or flange-back contact. This is the problem that generally limits wheel and rail life.

The application of oil or grease lubrication to the rail side or wheel flange has a double effect on the wear taking place. Firstly the lubricant lowers the effective coefficient of friction and therefore generally reduces the actual forces in the contact. Secondly it alters the mechanism by which wear takes place and this has a much more dramatic effect such that, as long as the lubricant film lasts, the wear can be considered almost negligible.

Water is partially effective as a lubricant. It can help to reduce forces but does not have the same dramatic effect as oil or grease on the wear mechanism.

On very sharp curves the use of flange, or high rail, lubrication does not necessarily reduce the gauge spreading forces as these are controlled by the friction force on the top of the low rail, an area that is not lubricated.

The large forces generated on sharp curves can also give rise to considerable unpleasant noise of flange or flange-back squeal. This can also be considerably alleviated by the application of lubrication.

Transitions and track twist

So far this section has concentrated on vehicle behaviour on steady curves. The situation on transitions is more complicated, particularly in the area of track twist and the resulting load transfer between the wheels of a vehicle and this can result in an increased derailment risk. In general, vehicles are designed and tested to be able to tolerate a specified amount of track twist within a limitation on the permitted wheel unloading.

The most critical situation for most vehicles is the run-off transition from a canted curve. In this situation the leading outer wheel of the vehicle is unloaded because of the track twist inherent in the transition whilst the leading inner wheel is overloaded and hence generates a particularly high friction force trying to push the vehicle off the track. This can in some circumstances lead to flange climbing derailments, particularly if rail irregularities or dips occur in such a position.

One particular problem that is worth consideration is the limits placed on rail side wear on curves. In general the effect of rail wear is to increase the effective gauge but not otherwise to have much effect on the vehicle behaviour until very large amounts of wear take place. However, if a curve needs to be rerailed for reasons of sidewear then considerable attention has to be paid to the transition from worn rail to new rail. If this is not adequately smooth then the change in rail shape provides an irregularity which will encourage flange climbing derailment.

Check rails

Check rails, or guard rails, are frequently used on sharp curves to reduce derailment risk and attempt to limit wear. The installation of a check rail, assuming it is correctly positioned relative to the high rail, distributes between the high rail flange and the low rail flange-back the force that would otherwise be taken wholly on the flange. In general the check rail and high rail wear until they are correctly matched and the forces distributed evenly between the two contacts. The wear is therefore distributed so that the use of check rails does prolong high rail life but the overall wear is generally little changed.

Check rails do have a significant effect on the risk of flange climbing derailment as the likelihood of the flange-back climbing the check rail as the flange climbs the high rail is much reduced.

Curved track ride

As mentioned earlier, the vehicle is more sensitive to lateral track irregularities on curves than on straight track as the wheelsets are more likely to be in flange contact with the rails. For passenger vehicles, investigation has shown that passengers are more sensitive to peak accelerations when they are superimposed on a steady state acceleration. This has the effect of making passengers particularly sensitive to lateral jolts on cant deficiency curves.

Figure 10 shows the relationships derived between dissatisfaction both for seated and for standing passengers and the steady state and peak lateral accelerations imposed. The message for track maintenance is that track on curves should be maintained to a better standard than on straights, both to reduce track forces from impacts and to maintain passenger comfort.

Conclusions

It is clear that vehicle–track interaction is important in a wide range of ways. On straight track, irregularities in top can create damaging track forces

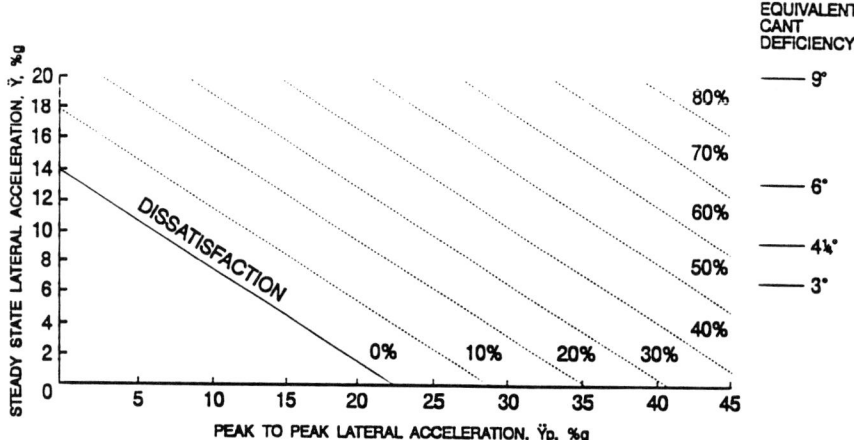

STANDING PASSENGER DISSATISFACTION FOR IRREGULARITIES

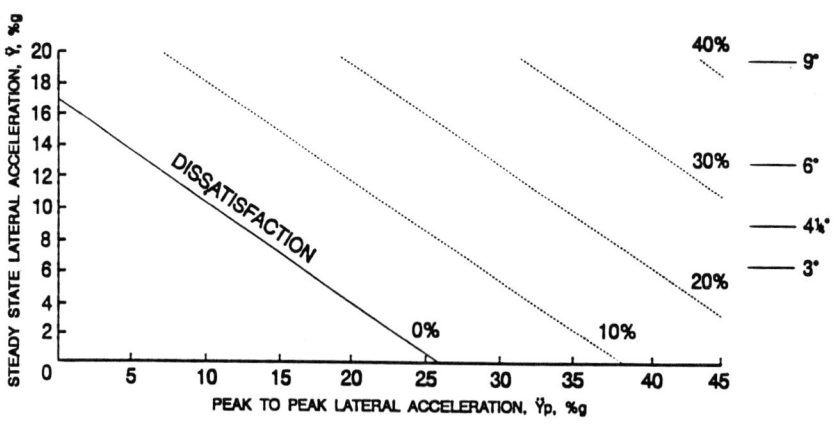

SEATED PASSENGER DISSATISFACTION FOR IRREGULARITIES

Fig. 10. Passenger dissatisfaction relationships

as well as causing unsatisfactory ride, while vehicle defects such as wheelflats can cause extremes of high frequency force which must be allowed for in track design. Rail head profiles can significantly affect vehicle stability at higher speeds.

Curving behaviour is complex — in addition to cant deficiency forces, wheelset/bogie internal forces are generated which encourage wear and the risk of derailment. Irregularities in curves are more problematical than on straight track, they are more difficult to remove and considerably more uncomfortable for passengers.

Modern computer techniques now allow the detailed investigation of most of these areas and studies can be undertaken of the sensitivity to individual parameters of the forces, wear etc.. Such methods are well worth considering in particular cases where economic or technical options have to be assessed and decisions made.

Acknowledgements

A summary of this kind covers the work of a large number of people over many years all of whom have made contributions to the understanding of the subject. Thanks are also due to the British Railways Board and the Director of Research for permission to publish this Paper.

2. Standards for track components

G. G. LEEVES, Group Technical Director, Pandrol International
Limited

Introduction

The performance of each component of the track structure has an influence upon the behaviour of the whole, but is itself influenced not only by
those other components but also by the quality and methods of maintenance
employed. Only rail can be looked at in isolation where the section is
determined by the axle load to be carried and choice is limited to the grade
and treatment of the steel to be employed and, to a lesser extent, dimensional
accuracy and particularly straightness. Of the other components, the choice
of sleeper has the greatest influence upon the behaviour of the total track
structure and will be considered first.

The sleeper

The choice is between steel, timber and concrete, each having its own
physical characteristics and maintenance requirements

Steel sleepers

Steel sleepers are manufactured to well-defined national standards like
British Standard 500 or the complementary UIC Standard 860-0. These
standards lay down chemical and physical requirements of the steel and the
sections from which sleepers shall be formed, to which some major steel
producers have added sections of their own designs. For the user, choice is
limited to details like the inclusion of rail inclination, the design of the spade
ends and the length. Physical tests on the raw material and dimension checks
on the finished product are normally the responsibility of the manufacturer,
overseen as desired by the customer.

Because a steel sleeper is in direct contact with the supporting ballast and
has no damping properties, all dynamic force excited in the rail is transmitted, together with any resonant forces generated within the sleeper, directly
into the ballast. The extent of these forces is dependent on the unspring
masses on the passing axles and the surface quality of the wheels and rails.
In very general terms, speeds in excess of the order of 100 km/h are difficult
to accommodate on steel sleepered track. On typical narrow gauge (1067 mm
and 1000 mm) track where steel sleepers have been widely used, limitations

of the order of 80 km/h for passenger stock and 80 km/h for freight stock are normal (these restrictions are not, of course, only a function of the sleeper type).

The section from which the steel sleepers is formed must be compatible with axle loading and typical figures are

Rail	Axle	Section	Thickness of plate at rail seat	Finished weight	
				Narrow	Standard
30 kg/m	15 t	21 kg/mm	10.3 mm	45 kg	60 kg
40 kg/m	18 t	30 kg/mm	11.9 mm	60 kg	75 kg
57 kg/m	25 t	32 kg/mm	12.7 mm	70 kg	85 kg

Timber sleepers

Timber sleepers do not have international standards covering their supply because of their great variability of performance across the wide species range used. There is a UIC Standard 863-0 which gives guidelines for species used in Europe but it is not more generally applicable. In practice, most railways have developed their own specifications to suit the performance of the timbers available to them. Preservation is to proprietary standards and detail is limited to the minimum volume of preservative absorbed during impregnation. The major cause of failure is rot caused by fungal spores entering the cove of the sleeper through cracks in the upper surface. The spore calories take several years to develop and effective treatment, fluoride or boric acid need to be applied 5-10 years after installation. This treatment could beneficially be specified as a requirement to maintain standards but not a purchasing standard.

Timber sleepers do not limit the capacity of a track structure in either axle load or speed, the sleeper size and/or spacing accommodating all conditions. Typical ranges of sections used are

Rail weight	Axle load	Speed max.	Section	Spacing
30 kg/m	15 t	60 km/h	200 mm x 125 mm	800 mm
40 kg/m	18 t	80 km/h	200 mm x 125 mm	750 mm
60 kg/m	25 t*	160 km/h*	250 mm x 125 mm	600-700 mm
		200 km/h	250 mm x 150 mm	500-600 mm

*Note: maximum axle load and maximum speed will not occur together.

Specific standards for sizes of acceptable cracks will depend upon timber species. UIC 836-0 limits length to 250 mm but not the number and requires split timbers to be braced by bolts, S clamps or other bracing methods. Gang rail plates are, arguably, the best system available. UIC 836-0 does not limit the location of cracks but they should not be permitted in or close to the line of any screw or spike securing devices. Incising should be specified for timber susceptible to cracking to inhibit the development of single cracks and encourage the development of a number of smaller ones. UIC 836-0 does not require sleepers to be protected with creosote or insecticide/fungicide agents. Where environmental regulations do not prohibit its use, pressure impregnation with a protective agent is very desirable.

Concrete sleepers

Concrete sleeper designs have been developed largely as co-operative ventures between railway administration and concrete product manufacturers. AREA have set out design guidelines and test procedures for two-block reinforced concrete and monoblock prestressed sleepers. National norms are usually specified for the concrete and its reinforcement.

Two-block sleepers

The AREA test requirements cover

- the rail seat positive bending movement
- the rail seat negative bending movement
- centre positive bending movement
- repeated load on inclined rail seat
- rail seat overload and ultimate load.

The criteria for acceptance in all but the ultimate load is the absence of cracking visible through a 5x magnifying glass.

Experience has shown that the two-block sleeper gives satisfactory performance for all speeds of traffic although they have not been proved satisfactory in heavy axle load conditions.

Monoblock sleepers

The AREA test requirements cover

- rail seat vertical load test
- centre negative bending movement
- centre positive bending movement
- repeated load on inclined rail seat
- bond development, tendon anchorage and ultimate load.

For production testing, British Rail specify only a rail seat positive bending moment test. Both AREA and British Rail specify a minimum cube strength at 28 days (7000 psi -48.26 N/mm^2). Sleepers would be required to meet national norms like British Standards BS5328 for the concrete and BS2691 for the reinforcement.

General comment

Traditionally concrete sleepers have been designed as static elements and dynamic forces have been accommodated within a live load multiplying factor (AREA recommend 200 %). Transient loads measured in track have been found to well exceed this figure (by at least a factor of 2) when dynamically stiff rail pads are used. Design needs to take account, therefore, of the type of rail pad to be used and the level of dynamic excitation to be expected.

Recent experience would suggest the introduction of limitations in the concrete specification regarding the maximum combined alkalinity of the aggregate and the cement to avoid the development of delayed eteringite formation.

Recent experience in the United States with the erosion of rail seat areas under high axle load, the development mode of which is still undetected, would indicate caution in the design of the rail pad and rail seat area of sleepers carrying 30t+ axle loads. There is no evidence of this problem occurring in European loading conditions.

The rail

Three main international standards, British, Continental (UIC) and American (AREA), dominate the supply of rails although there are also a number of national standards. Grades are usually defined by their tensile strengths which, under normal conditions are proportional to harness and wear resistance. Typical characteristics are

	Min. tensile strength	Hardness BHN
Standard carbon (0.4%-0.60%)	$710\,N/mm^2$	220
Wear resistant (A) (0.6%-0.8%)	$880\,N/mm^2$	265
Wear resistant (B) (0.5%-0.75%C) (1.30%-1.70%Mn)	$880\,N/mm^2$	265
High strength (0.72%-0.82%C) (0.85%-1.10%Mn)(0.95%-1.20%Cr)	$1080\,N/mm^2$	320
Heat treated	$1150\,N/mm^2$	340

Choice will depend on track conditions. In typical European conditions with 22t axle loads, typical usage is

Standard carbon	$710\,N/mm^2$	Tangent track and curves greater than 800 m R
Wear resistant quality	$880\,N/mm^2$	Tangent track and curves greater than 500 m R
Silico manganese wear resistant	$980\,N/mm^2$	Curves sharper than 300 m R

High strength chrome manganese	1080 N/mm^2	Curves sharper than 250 m R

In Africa on narrow gauge track and traffic levels of 6-10 mgrt/year, typical usage is

Standard carbon	720 N/mm^2	Tangent track and curves greater than 300 m R
Wear resistance quality	880 N/mm^2	Curves sharper than 300 m R

and with traffic levels less than 6 mgrt/year, standard carbon in all track in including curves of 100 m R.

The choice of rail quality is not only dependent on speed and axle loading. Factors like rail corrugation and the development of rail head defects like shelling can justify the use of wear resistant rail where side wear alone might not, and sidewear is very unpredictable with wide ranges of rates of wear in ostensibly similar operating conditions.

The rail pad

Rail pads are generally not used with steel sleepers unless electrical insulation is required and are not necessary with timber sleepers. They are, however, an essential constituent of concrete sleepers. Rubber pads are covered by UIC Standard 864-5 but pads are generally ordered against performance specifications provided either by the railway or the pad or fastening manufacturer.

In steel sleepers requiring electrical insulation, an elastomeric pad 4 mm - 6 mm thick is most appropriate. Durability is generally the most desired characteristic after electrical resistivity. Liquid cast polyurethanes probably provide the most durable pads and modified high density polyurethane the cheapest. Capital cost versus maintenance cost policies would drive the decision. The detailed specification for the selected pad material would depend upon the manufacturers published data and tests to prove its physical properties would be to the national norms. particular attentions being paid to temperature behaviour and compression set characteristics. Tests on the finished produce would probably be limited to hardness and dimensional tolerances. As the pad influences the overall behaviour of the rail securing system, it would naturally be included in the system tests.

With concrete sleepers the rail pad is very influential in both the performance of the sleeper and of the whole track structure. The concrete sleeper has critical resonances at its first and second symmetrical modes which can be readily excited by imperfections in the rail/wheel contact surfaces, especially if the sleeper is not well bedded into the ballast. Whilst good ballast support has a sharply damping effect, in practice transient strains can exceed the quasi-static strains (developed by a perfect wheel/rail interface) by a factor of four and dominate the force regime. There is no opportunity to absorb these forces in the rail seat area because there is insufficient space in which to provide an adequate volume of material to absorb the heat generated without seriously degrading it. It is possible to isolate 60%-70% of the dynamic forces from about 30 Hz to over 2 kHz which properly designed, highly resilient rail pads. The essential characteristic of such a pad is its dynamic resilience within. Natural rubber with a minimum of filler is the most reliable material, liquid cast urethanes are the most rugged and suitable for sharply curved track. Typical attenuation and secant moduli figures for popular proprietary pads are

Material	Overall thickness	Attenuation	Secant stiffness
Studded natural rubber	10 mm	60%	40 kN/mm
Plain rubber bonded cork	10 mm	25%	140 kN/mm
Grooved synthetic rubber	10 mm	36%	180 kN/mm
Studded natural rubber*	6.5 mm	48%	90 kN/mm
Studded thermoplastic elastometer	6.5 mm	25%	140 kN/mm
Plain rubber bonded cork	5.0 mm	17%	350 kN/mm

Apart from the reduction of bending strain in the sleeper, the effect is also seen in a reduction of ballast disturbance. Theoretically, the contact stress at the wheel/rail interface is a function of track modulus and the reduction obtained with highly resilient pads should substantially reduce the contact stress. This effect remains to be proven but early results on one test site give encouraging indications that the development of corrugation is retarded.

Rail pads are normally supplied against a peformance specification with material specifications largely following the manufacturers figures. For a 10 mm thick studded natural rubber pad, typical limitations could be

Tensile strength	min 17 MPa
Elongation	min 300% at break
Electrical resistivity	min 1×10^8 Ohm cm
Hardness	$70 +^5$ IRHD
Compression set	
22 hours at 70°C	max 30%
70 hours at 23°C	max 20%
Dimensions	as specified in drawing
Attenuation	min 50%

National norms and standards would be applied to the physical tests. Attentuation would be measured using a Battelle type drop weight test.

For a plain surfaced thermoplastic pad, typical corresponding limitations could be

Material	A proprietary material(s) would be named
Physical tests	Melt flow index
Density	To manufacturers specifications
Hardness	To manufacturers specifications
Tensile strength	To manufacturers specifications
Elongation at break	To manufacturers specifications
Dimensions	As specified in drawing

Attenuation would not be specified as it would be almost zero at 6.5 mm thickness and 2-3% at the very most at 10 mm.

Shaped thermoplastic, grooved rubber and rubber bonded cork pads would have specifications similar to the studded rubber pad but with figures corresponding to their material characteristics.

In addition to these characteristics, some form of static modulus is usually specified. This can vary from upper and lower exceedance curves to a secant modulus between two specified loads, usually the lower representing the permanent clamping load of the fastening and the upper the extent of the load expected in service. Conventionally these limits are 20 kN and 100 kN.

Insulators

Most fastening systems incorporate one or more electrical insulating members other than the pad. The components are invariably load bearing and the most commonly used material is nylon, with or without reinforcement. The items are usually proprietary to the fastenings system and supplied against manufacturers' specifications. The physical tests on the material will be to national standards and cover the manufacturers' specifications for such characteristics as density, melt point, electrical resistivity, tensile strength and hardness. Finish and dimensional accuracy checks with possible sectioning to examine the internal structure for porosity to a foredetermined sampling plan would form part of the acceptance test.

Rail fastenings
Non-resilient rail fastenings

This category includes all the earliest rail fastening systems where the rail is rigidly secured to the sleeper. All forces generated in the rail are transmitted through the fastening system to the sleeper. The fastening, the sleeper and the ballast are all subjected to vibrational forces and potential fatigue problems, making them unsuitable for traffic over 45 miles/h and undesirable except for very lightly used lines. Maintenance will be more expensive than with resilient fastenings but significantly cheaper in capital cost.

Resilient fastenings

This category includes all modern fastenings, most of which are patented and the proprietary products of specialist manufacturers. These fastening systems include a resilient member securing the rail which maintains contact when the rail moves relative to the sleeper. This permits the use of a resilient member between the rail and the sleeper and the opportunity to isolate some of the dynamic force generated in the rail from the sleeper and the ballast.

Fastening specifications

The materials used for fastenings are usually covered by national norms. No such norms cover fastening assemblies which are conventionally specified either in terms of performance or in accordance with the manufacturers' specifications. AREA, in Chapter 10 of their manual, specify a range of tests which combine the performance of the sleeper and the fastening and include: repeated inclined load; rail uplift; rail rollover; and longitudinal restraint.

The criteria for acceptance of all but the last item is the survival of the components. In the lateral restraint test, a specified resistance before run through (1t/sleeper/rail). These tests do not accurately simulate actual conditions, largely because the impact forces which generate the large, but damaging transient forces and the large resonant forces cannot be reproduced in the laboratory, except as single incidents. Conventional servo hydraulic equipment is limited to about 20 Hz and at that frequency are running out of amplitude. In practice the only quantitative tests of value have to be undertaken in actual track conditions.

All modern resilient fastenings are suitable for high speeds and all but the most severe heavy haul conditions. Their relative merits lie mainly within their practical characteristics of ease of installation and removal, their adaptability, their maintenance needs, their durability and their life-cycle costs.

Ballast

Ballast is often the most expensive element in the track structure but it is probably the least researched and understood. Provision will, in many cases, be dictated by availability by when choice is available, characteristics of hardness, density and resistance to abrasion are of greatest importance. Typically the Los Angeles abrasion test is specified with a maximum figure of 25%-30%. British Rail have developed their own abrasion test where a sample is mixed with its own weight of water and tumbled. There are no national norms for ballast. Ballast sizes from 65 mm-20 mm and 50 mm-25 mm are typical. Maximum density by properly grading the ballast has been generally seen as an advantage.

RAILWAY ASSETS

(1988 or latest available year)

RAILWAY:	Line Km	Total Staff	Diesel Locos	Elec Locos	Freight Wagons	MU Fleet	Coaches
LATIN AMERICA:							
Antofagasta & Bolivia	750	622	31		2,535		3
Argentina	34,140	98,000	990		40,300		2,500
Bolivia	3,652	6,761	67		2,160	37	129
Brazil – – FEPASA	5,035	19,000	360	140	12,000	140	260
Brazil – – RFFSA	22,067	62,000	1,515	40	45,000	63	1,247
Chile	6,969	7,150	169	100	6,323	59	437
Colombia	2,616	7,730	158		3,998	19	124
Costa Rica	480	2,700	70	21	1,455		
Cuba	4,161	34,500	418	12	10,387	90	400
Mexico	15,825	81,248	1,742		47,000		1,202
Peru	1,667	5,820					
Uruguay	2,991	5,767	59		2,545	7	72
AFRICA (Sub Saharan):							
Cameroun	1,104	6,244	92		1,801	4	104
Congo – – CFCO	610	5,849	54		1,800		120
Cote D'Ivoire	1,167	5,600	38		1,500	16	142
Ethiopia	781	2,500	29		630	32	37
Ghana	950	7,800	65		1,900		130
Kenya	2,084	21,650	209		6,500		547
Malawi	789	4,512	32		814	1	32
Mali	642	2,738	34			9	54
Nigeria	3,512	24,971	209		7,000		663
Senegal	904	2,027	29		810	8	97
South Africa	23,507	122,000	1,561	2,350	167,181	4,600	7,708
Sudan	4,764	32,000	137		6,700		4,251
Tanzania	2,584	17,000	84		4,110		125
TAZARA	1,860	3,600	80		1,800		100
Uganda	1,232	6,300	60		1,280		98
Zaire	4,511	22,253	124	45	5,438	19	251
Zambia	1,278	8,700	70		6,700		84
Zimbabwe	2,759	17,293	307	30	12,418		334
EMENA:							
Algeria	4,135	17,186	201	24	13,300	28	737
Morocco	1,893	13,440	128	100	8,514	16	591
Tunisia	1,931	9,124	198	6	5,176	45	348
Egypt	5,870	82,000	780		14,000		3,377
Portugal	3,608	22,064	267	48	5,322	272	560
Albania	672	670	65		1,500		68
Bulgaria	4,300	67,111					
Czechoslovakia	13,103	236,109	4,446	1,843	134,051	1,178	8,888
Hungary	7,614	129,038	1,095	468	66,672	320	4,129
Poland	26,545	360,015	2,590	1,941	152,274	1,000	6,300
Romania	11,110	176,000	2,000		126,000		
Yugoslavia	9,349	153,038	832	504	52,386	479	3,784
Turkey	8,164	59,000	650	18	20,255	107	1,344
Pakistan	8,774	132,314	566	29	36,000	14	2,600
ASIA:							
Bangladesh	2,818	54,000	290				1,389
Burma (Myanmar)	3,185	27,500	272		8,141		1,246
China	52,767	3,293,000	4,836	1,197	340,300		24,917
India	61,976	1,624,100	3,454	1,533	345,821	3,052	27,738
Indonesia	6,458	51,000	518				1,200
Republic of Korea	3,150	38,674	483	93	15,311	681	2,157
Malaysia	1,668	9,300	90				249
Sri Lanka	1,453	26,000	180			46	1,283
Thailand	3,818	24,926	279		8,689	186	1,135
Vietnam	2,504	64,000	312		5,358		1,000
OTHER COUNTRIES:							
Austria	5,783	67,061	488	720	31,000	313	2,800
Belgium	3,568	53,000	638	354	34,000	688	2,044
Finland	5,900	24,000	382	110	18,600	199	763
France	34,563	213,214	1,976	2,290	169,000	5,680	10,300
Greece	2,479	14,000	214		10,500	119	467
Ireland	1,944	14,000	126		1,740	78	238
Israel	869		40		1,289		73
Italy	16,002	213,344	445	1,982	110,082	1,625	14,556
Japan	20,934	198,164	940	1,100	29,000	20,700	2,726
Netherlands	2,809	27,000	406	148	8,100	699	480
Spain	12,686	60,000	756	623	40,000	719	2,043
Sweden	11,194	34,000	532	694	35,000	30	1,477
United Kingdom	16,599	149,500	1,542	184	46,000	3,321	4,018
West Germany	27,284	251,344	2,928	2,612	220,401	2,143	14,971
USSR	147,300	1,814,000	20,362	12,600	1,690,000	9,900	56,000
Australia:ANR	7,147		214		6,605	14	217
Australia:SRA N.S. Wales	10,800	34,242	649		9,209	2,068	43
New Zealand	4,029	12,560	371	20	14,770	167	104
Canada: Via Rail	18,500	6,873				296	594
Canada:Canadian National	50,708	42,000	1,600		76,000		172
Canada:Canadian Pacific	33,458	25,000	1,000		38,000		
USA:Amtrak	38,710	24,000	237	59		43	1,810
USA:Commuter Railways	7,690	25,000				1,920	2,000
USA:Burlington Northern	37,865	32,000	2,350		56,000		
USA:Conrail	21,518	30,487	2,323		76,177		
USA:Denver & Rio Grande	3,624	2,000	310		10,000		
USA:Florida East Coast	785	1,000	71		2,800		
USA:All Class I Railways	205,734	227,548	19,364		652,123		

35

RAILWAY OUTPUTS
(1988 or latest available year)

RAILWAY:	Freight Tonnes (000,000)	Tonne–KM 1988 (000,000)	Passengers 1988 (000,000)	Pass–Km (000,000)	TU/ Employee (000)	TU/Km (000,000)	P–KM To Total TU (%)	Freight Haul (Km)	Passenger Trip (Km)
LATIN AMERICA:									
Antofagasta & Bolivia	1.5	385			618	0.51		570	
Argentina	14.9	8,257	352,000	12,400	212	0.61	60.2	570	35
Bolivia	1.0	500	1,390	500	149	0.27	49.8	514	359
Brazil –– FEPASA	22.0	7,200	93,000	2,900	510	2.01	29.0	327	31
Brazil –– RFFSA	82.0	37,000	41,000	1,490	610	1.74	3.0	453	36
Chile	6.6	1,700	7,200	1,100	400	0.40	40.3	263	162
Colombia	1.0	562	1,400	177	100	0.28	24.0	580	125
Costa Rica	1.0	79	1,922	72	55	0.31	47.0	90	38
Cuba	14.0	2,155	23,000	2,180	135	1.04	50.3	159	93
Mexico	57.4	41,177	18,487	5,619	576	2.96	12.0	718	304
Peru	2.0	620	3,400	500	180	0.67	42.8	260	137
Uruguay	1.0	213	2,240	141	37	0.12	39.8	217	63
AFRICA (Sub Saharan):–									
Cameroun	1.4	594	2,414	470	167	0.96	44.1	432	190
Congo––CFCO	1.1	448	2,200	400	145	1.39	47.2	405	175
Cote D'Ivoire	0.7	499	1,600	500	170	0.86	49.5	669	310
Ethiopia		144	900	220	143	0.47	60.4	473	244
Ghana	0.7	126	3,610	353	47	0.50	73.7	169	98
Kenya	3.2	1,700	3,200	700	109	1.15	29.2	552	220
Malawi	0.3	71	1,664	112	40	0.23	61.2	204	67
Mali	0.4	227	800	177	147	0.63	43.8	505	221
Nigeria	0.3	207	4,276	855	40	0.30	80.5	634	200
Senegal	3.2	460	940	89	200	0.61	16.0	144	90
South Africa	163.7	87,080	578,015	13,874	827	4.29	13.7	532	24
Sudan	0.6	680	2,000	800	46	0.31	54.1	907	396
Tanzania	1.0	680	2,200	769	104	0.56	52.9	768	393
TAZARA	1.2	1,400	1,330	400	260	0.97	21.4	1,193	298
Uganda	0.4	78	1,700	212	45	0.24	73.1	207	120
Zaire	3.9	1,859	1,000	389	101	0.50	17.3	477	389
Zambia	5.8	1,300	1,800	420	202	1.35	23.7	233	224
Zimbabwe	13.2	5,568	2,740	870	372	2.33	13.5	421	318
EMENA:–									
Algeria	13.0	2,799	44,900	2,439	305	1.27	46.6	215	54
Morocco	32.3	5,605	11,556	2,092	572	4.07	27.2	174	181
Tunisia	11.4	2,144	28,190	1,014	346	1.64	32.1	188	36
Egypt	9.0	3,000	481,000	23,900	320	4.58	88.8	330	45
Portugal	6.0	1,597	230,500	6,036	346	2.12	79.1	266	26
Albania	8.1	600	8,800	700	162	1.93	57.2	75	64
Bulgaria		18,100	107,800	8,143	121	6.10	31.0	218	76
Czechoslovakia	295.1	75,290	415,366	19,408	401	7.23	20.5	255	47
Hungary	111.4	20,737	223,700	11,395	249	4.22	35.5	186	51
Poland	417.7	120,671	983,800	52,134	480	6.51	30.2	289	53
Romania		78,000	347,900	23,220	580	9.11	22.9	286	67
Yugoslavia	83.6	25,414	115,700	11,449	241	3.94	31.1	304	99
Turkey	14.3	8,006	135,700	6,708	227	1.80	45.6	560	49
Pakistan	9.3	8,000	81,000	18,000	200	2.96	69.8	820	228
ASIA:–									
Bangladesh	2.5	678	53,000	5,052	120	2.03	88.2	269	95
Burma (Myanmar)	1.6	386	60,859	4,172	144	1.43	91.5	241	69
China	1,406.0	986,020	1,216,000	325,731	398	24.86	24.8	701	268
India	329.5	230,131	3,792,100	263,731	304	7.97	53.4	698	70
Indonesia	7.9	1,600	48,300	7,300	172	1.38	82.0	203	151
Republic of Korea	60.7	13,784	564,200	25,978	1,028	12.62	65.3	227	46
Malaysia	3.1	1,100	7,400	1,520	272	1.57	56.0	364	205
Sri Lanka	1.5	195	59,000	1,900	80	1.44	90.6	130	33
Thailand	6.2	2,867	82,706	10,301	528	3.45	78.2	461	125
Vietnam	3.9	1,016	17,800	3,506	71	1.81	77.5	258	197
OTHER COUNTRIES:–									
Austria	58.0	11,400	160,000	8,400	279	3.42	40.1	202	90
Belgium	76.0	7,300	142,000	6,270	256	3.80	46.3	95	44
Finland	30.0	7,500	41,400	2,670	423	1.72	26.2	250	65
France	143.3	51,527	801,100	63,057	537	3.32	55.0	360	79
Greece	3.9	584	11,700	1,900	200	1.00	77.2	150	164
Ireland	3.0	570	24,700	1,190	120	0.91	67.7	188	48
Israel	6.6	1,035	2,495	161		1.38	13.5	157	65
Italy	58.2	19,567	410,000	43,343	295	3.93	68.9	336	106
Japan	55.7	23,000	7,761,000	217,587	1,214	11.49	90.4	413	28
Netherlands	18.6	2,900	221,900	9,300	450	4.34	75.8	161	42
Spain	34.7	11,000	190,300	15,300	430	2.07	58.3	316	81
Sweden	51.7	17,000	71,000	6,100	663	2.06	26.0	331	85
United Kingdom	149.5	18,104	763,700	34,315	350	3.16	65.5	121	45
West Germany	273.9	58,972	1,025,900	40,959	398	3.66	41.0	215	40
USSR	4,116.0	3,924,800	4,396,000	413,800	2,349	29.41	9.5	954	94
Australia:ANR	15.3								
Australia:SRA N.S. Wales	50.2	13,553	249,296	4,275	521	1.65	24.0	270	17
New Zealand	8.9	2,924	15,000		233	0.73		328	
Canada: Via Rail			6,415	2,303	335	0.12	1.1		359
Canada:Canadian National	108.4	128,000			2,780	2.52		1,180	
Canada:Canadian Pacific	82.0	84,000			3,200	2.51		1,030	
USA:Amtrak			21,496	9,463	394	0.24	0.6		440
USA:Commuter Railways			319,200	11,134	445	1.45			
USA:Burlington Northern	243.0	300,000			9,200	7.92		1,250	
USA:Conrail	174.6	127,330			4,177	5.92		729	
USA:Denver & Rio Grande	21.1	16,000			7,900	4.42		758	
USA:Florida East Coast	14.2	6,000			5,900	7.64		438	
USA:All Class I Railways	1,297.2	1,477,488			6,264	7.18		1,139	

Discussion on Papers 1 and 2

F. I. MAU, Vice-President Operations, BHP Rail Products (Canada) Ltd
Standards appropriate to developing countries' railways need to be carefully considered in the light of available maintenance resources. Reference is made in Paper 2 and several other papers to the recommendation generally considered to hold true that concrete sleepered/continuously welded rail track should be considered the optimum.

However, I suggest careful consideration needs to be given to the implications on maintenance. I note in this context the need to place rail at the appropriate stress-free temperature, and maintain it there. This is a significant task in North America; in a developing country it may not be possible.

H. M. AHMED, former General Manager, Sudan Railways
Bridges and structures: Inspection of bridges and culverts is best entrusted to the permanent way section because they are on site every day. Some of the defects discovered by permanent way gangs are

(a) The bombing of the Lol Bridge was first reported by permanent way gangs
(b) some steel angles bracings which were unbolted from the bottom boom of a steel bridge were discovered by the Permanent Way Inspector during his patrol
(c) the permanent way section notice changes in the course of streams and can advise on new locations for bridges or culverts
(d) the regular cleaning of culverts, especially in relation to moving sand and storms should be undertaken by the permanent way section. The bridge section do not have sufficient staff to have enough people on site to undertake this work in addition to their other duties.

However, it is essential that permanent way gangs are taught the basics about bridges in order to avoid errors if they are undertaking work of this nature.

M. F. SAVASTANO, Sir Alexander Gibb & Partners Ltd, London
Ms Eickhoff discussed the kinematic behaviour of wheelsets in relation to the cone angle and mentioned that an equivalent conicity can be determined.

DISCUSSION

On British Rail the cone angle of 1 in 20 is compatible with the rail head planing angle and rail inclination; on overseas railways however, rail inclination may be 1 in 40 for example. What are the effects of varying these criteria and is there an optimum combination of coning angle, head planing angle and rail inclination which gives a longer life to the rail and reduces maintenance costs?

H. M. AHMED, former General Manager, Sudan Railways

Further to his study on the standards of sleepers and comparison between the three different types could Mr G. G. Leeves compare the standards, quality and cost of the three methods. Which is best? Could he arrange the three types in order of quality, technical characteristics and cost.

For example in Sudan we have a lot of timber resources, (we have forests with good quality hardwood) but compared to imported wood it proved to be more expensive. Nor do local factories have the capacity to supply the full requirements of annual maintenanace and track renewals.

We also in certain cases prefer steel sleepers to timber sleepers as the lifetime is more than three times that of timber sleepers. This is because of the climate in Sudan and the effects of termites and other insects which lead to the deterioration and short lifetime (the average lifetime of timber sleepers is 12 years). In my opinion, generally the most economically viable is steel, then timber and lastly concrete. (NB We dont use concrete sleepers although we have the basic resources (gravel) because we inherited an earth embankment and we dont have mechanized maintenance. (The concrete sleepers need mechanized handling facilities. It can't be done the manual way.)

So I would be pleased for an extension of the sudy to include a world-wide evaluation of the three types taking into consideration the different conditions of each railway and of each situation. The arrangement may differ in certain occasions or certain countries but the principle is the same.

H. M. AHMED, former General Manager, Sudan Railways

Regarding the effect of the irregularities of track on rolling stock, it is not fully agreed which came first, but a lot of dispute is going on. Mechanical engineers always complain about the effect of track irregularities on the rolling stock, locomotives, wagons, passenger coaches etc. Could Ms Eickhoff comment on the percentage of the effect of track irregularities on rolling stock? The rolling stock also has its problems, its movement, running and operations are affected by fatigue. It also affects the track and causes fatigue and irregularities increasing the wear and tear in other areas and occasions.

F. I. MAU, Vice-President Operations, BHP Rail Products (Canada)Ltd

Vertical, lateral and V/L ratios in North America are commonly considerably higher than those noted in Ms Eickhoff's paper. Vertical wheel loads of 180 kN are common, lateral loads of 90 kN are also the norm. However, much higher figures occur and when dynamic loading from dipped welds or wheel flats, for example, are added, Union Pacific Railroad have detected vertical loads above 450 kN. L/V ratios are commonly 0.5 on the inner (low) rail, 0.7 on the outer (high) rail. These all raised questions about track components capable of providing cost-effective support for such loads.

The loads mentioned affect the vehicle as well as the track. Damage to bearings for example, must be recognised in determining maintenance standards.

Has the BR analytical model been validated by track measurement? How have longitudinal wheel forces been measured?

M. F. SAVASTANO, Sir Alexander Gibb & Partners Ltd, London

Ms Eickhoff mentioned that modern vehicle design is adequate down to a radius of approximately 600 m, special steering mechanisms being required for sharper curves together with improved track design.

With curves sharper than 200 m radius gauge is widened by up to 19 mm. Within switch and crossing work, however, this is not the case on British Rail, although gauge widening is provided in many overseas designs.

There is a considerable amount of switch and crossing track within BR and in private sidings using turnouts with radii sharper than 200 m. What are Ms Eickhoff's views on gauge widening of turnouts to alleviate lateral force effects and relieve stresses on fastenings, caused by three-axled loco-motives?

J. K. YATES, Manager Products Process Department, British Steel, Track Products, Cumbria, UK

Mr Leeves comments that steel sleepers are restricted to speeds of less than 80 km/h. This may be true in Africa, with bare steel rail on bare steel sleepers, but is not an accepted ceiling. Increasing numbers of steel sleepers are being used in secondary tracks by BR with passenger speeds of 140 km/h and freight axle loads of 25 t, using the same resilient pads and inserts as are equally necessary for concrete sleepers.

A. WILSON, Consultant, Clyde Engineering, Sydney, Australia

Desirable standards: My company manufactures diesel electric and electric locomotivess and suburban cars. In Australia, diesel electric locomotives are built under licence from North American manufacturers, but most locomotives have to be re-designed to fit under a loading gauge of only 4270 mm, and to have a lower axle load than that used in North America. The Australian manufacturer has to squeeze many desirable features into a smaller volume and a lower overall mass.

Railways world-wide seek to reduce their fuel consumption and there is intense competition by the locomotive manufacturers to develop new technology, which will increase power outputs, adhesion levels, fuel efficiency and the reliability so essential for power by the hour leasing contracts.

Progress is rapid and nowadays three or four older locomotives can be replaced by two new ones, using less fuel to haul a train of the same mass.

But the trend is for the new locomotives to become heavier than the older ones. Other requirements add more mass, for example

(a) environmental legislation requires larger silencers and more effective sound proofing

(b) train crews require better facilities, e.g. air conditioning, isolated cabs to reduce vibration and stronger collision protection

(c) cabs at both ends, especially for low traffic lines

(d) although more efficient, a more powerful engine requires a larger fuel tank

(e) dynamic braking is a very desirable feature.

The result is that a modern state of the art 16 cylinder 2863 kW locomotive, with these features, will have an axle load of 22 t, with a trend towards higher axle laods.

On heavy gross tonnage lines, the railway can justify heavy track suitable for axle loads up to 30 t or more.

But on light traffic lines, revenues are greatly reduced and it is difficult to justify large expenditure on upgrading the track system.

The message from the mechanical engineer's viewpoint is that a reduction in axle loads below 22 t, even by only a few tonnes, will result in a heavily compromised locomotive solution, such as a half powered locomotive, with only a small reduciton in the capital cost.

G. G. LEEVES, Author

Mr Ahmed raises the question of choice between timber, concrete and steel sleepers which is a subject in itself and Mr Mau rightly questions the premise that concrete sleepered/continuously welded track should be considered as optimum. For minimum dynamic conditions, timber sleepers induce the least ballast disturbance and the quietest track, (noise is, after all, the manifestation of vibration and resonance). Timber is, however, a natural material which can be very variable in quality and durability and life cycle costing will depend upon its continuing ability to restrain the loosening of the rail securing devices, to maintain its physical integrity (without undue cracking or insectival or fungal attack and degradation) and to deformation in the rail seat area. Cost will depend on availability (whether it occurs in stands of similar species), on location and ease of transportation, on the existence of adequate milling facilities and on competitive demand for alternate usage (cutting straight sleepers from a log frequently gives fairly poor usage and a lot of offcuts which may have little commercial value). In general, timber sleepers can be used in all locations except those where insectival attack cannot be resisted. They provide the most stable track in poor formation conditions and require less ballast than concrete sleepers but more than steel. They are the most tolerant of adverse ballast support conditions.

Steel sleepers provide very stable track because of their shape and very good gauge control. They are, dynamically, very lively and retention of ballast compaction in areas of excitation is difficult and extremely time consuming. Mr Yates comments on the use of steel sleepers on British Rail on tracks carrying traffic at 140 km/h. I am not aware of the conditions but I would expect the line, level and rail surfaces to be of a particulary high standard if significant ballast maintenance problems were not to be experienced. The conference subject was related to railways in developing countries where climatic, formation, economic and traffic conditions might not justify the track quality and train speeds of British Rail. My experience with ballast maintenance under steel sleepers leads me to question the efficacy of mechanized maintenance. Steel sleepers are, however, ideally suited to manual or mechanically assisted maintenance. They are immensely rugged and durable but are capable of being handled by one person and can survive with little or no ballast.

Concrete sleepers provide heavy high modulus track but they have physical characteristics which make them very vulnerable to dynamic forces. The correct design of the rail support system and the quality of the ballast support are critical to the maximum isolation of damaging high frequency forces from the sleeper and a maximum damping effect from the ballast. They are too heavy for manual maintenance except for the treatment of individual sleepers and require the use of heavy and relatively complex equipment and its associated maintenance support. Concrete sleepers require significantly more support ballast than either timber or steel and because of their relatively poor frictional restraint and their tendency to resonate, a wider ballast shoulder and good crib ballast are needed to maintain their locational stability. Because of the lack of damping within the sleeper, the ballast is subjected to high crushing and dynamic forces and for an economic life, good hard, attrition resistant ballast is essential.

Mr Ahmed notes that in his railway's specific conditions, steel sleepers provide the most practical and economic track. With his climatic conditions and the manual maintenance methods employed, the choice of steel sleepers is almost certainly correct. On another system where labour costs motivate the maximum use of mechanization and skilled personnel are available adequately to maintain complex track maintenance machines, timber or concrete sleepers would almost certainly provide the most economic track structure. In each case the choice will depend upon the circumstances and one cannot generalize on what is best or indeed what is the most economic.

B. EICKHOFF, Author

In response to M. S. Savastano, the most important consideration in selecting wheel coning angle, head planing angle and rail inclination is that the wheel and rail should be compatible, so that a 1 in 20 inclined rail should be used with wheels based on a 1 in 20 and a 1 in 40 inclined rail with a 1 in 40 based wheel. If the wheel and rail are not compatible, but are based on different angles, then either very high or very low values of equivalent conicity can be generated leading to dynamic problems. As long as the wheel and rail are based on the same angle then the actual value of the angle is of secondary importance.

In response to H. M. Ahmed, the magnitude of wheel-rail forces generated is dependent on the level of irregularity in the track and on the vehicle suspension, operating conditions, speed etc, and the forces then feed into both the vehicle and the track. A given level of track force can be generated by a number of different vehicle-track roughness combinations. A more sophisticated vehicle might produce the same forces on rougher track as did a more basic vehicle on smoother track. The appropriate solution has to be found for any particular set of circumstances by consideration of the costs of the system as a whole and assessment of the options.

In response to F. I. Mau, a considerable number of track tests have been carried out using freight vehicles, passenger vehicles and locomotives to validate our theoretical methods. These tests have looked at behaviour on curved track and also on straight track and through irregularities such as switch and crossing work. Good agreement has been obtained between theory and experiment. Wheel– rail forces have been measured in all three directions (vertically, laterally and longitudinally) using load measuring wheels with strain gauged spokes.

In response to M. F. Savastano, the original purpose of gauge widening was to alleviate the constrained (or jammed) position that long wheelbase vehicles with rigidly connected multiple axles (e.g. steam locomotives) could reach on sharp curves. With more modern three-axle bogies the amount of lateral freedom between the wheelsets and the bogie frames allows the vehicle to negotiate very much sharper curves without this problem occurring. A study of the use of gauge widening on British Rail concluded that, with today's rolling stock, the effect of gauge widening was frequently to increase (rather than decrease) gauge spreading forces as the additional clearance allowed the bogie to rotate further and generate a larger angle of attack at the leading wheelset.

3. Maintenance tolerances — desirable and minimum values

P. FORBES, BSc, MICE, Regional Railways Civil Engineer, British Rail

Introduction

Desirable tolerances are those that give the most cost-effective whole railway solution. Minimum tolerances are those that give a safe railway.

It is important and sometimes difficult to differentiate between the two. Clearly the desirable tolerance can never fall below the minimum tolerance and the prime aim of permanent way Engineers is to make sure that the trains stay on the tracks, therefore the minimum tolerances are always set to achieve this aim.

However, this may not give a comfortable ride to passengers or give a cost effective solution, therefore much time has been put in on many railway networks to determine what is a desirable tolerance that is economic to achieve and at the same time gives the passenger the comfort he wishes.

There are a number of factors that have to be taken into account before formulating standard tolerances.

Factors for consideration

The original design

When considering the track and formation, whether it was laid in CWR or jointed track, whether the railway was built with straights and curves or with transitions between, whether enough detail was put into the formation and indeed the type of material are all factors which may have been decided upon at the time the railway was built. They all have to be taken into account when deciding upon maintenance standards and frequencies.

For example, the different approaches of the original railway engineers in Great Britain in building the Great Western Main Line and the East Coast Main Line was such that Brunel would not tolerate any level crossing whereas there were some five hundred on the East Coast Main Line. The introduction of faster trains, namely the HST 125, was first made on the Great Western Main Line some years ahead of other routes simply because of this design consideration which did not anticipate the trains that were eventually run on the routes.

Speed and tonnage of traffic

Both the speed and the tonnage have an effect on how quickly the track moves from its design tolerances. Heavy axles tend to hammer joints. Damage occurs at any discontinuity in the track and punishment of S & C work is a particular problem.

Speed has the effect of putting strain on the various components particularly on the high rail and can also push alignment out if the track is not firmly fixed. It can also be a major factor in the rate of wear.

Mix of traffic

If a route has a traffic pattern which means that all the vehicles are of a similar nature then problems tend to be of a specialist few, whilst if you are considering a mixed traffic railway where there may be light high speed passenger traffic and slower heavy-axled freight traffic then track problems can be of a great and varied nature.

Age of components

The age of the track has an effect on the tolerances that are acceptable. This is particularly true with metallic components which begin to suffer from fatigue failure and the forces placed upon these components have to be decreased or components have to be replaced.

British railways track tolerances

When considering existing British Railways track tolerances, there are a number of parameters which are measured today using the High Speed Track Recording Coach (HSTRC). Namely they are three metre twist, five metre twist, dynamic cross level, top left above datum, top left below datum, top right above datum, top right below datum, the alignment itself and also the gauge.

Twist

I will consider 3 m twist and 5 m twist together. The major factor affecting twist is the design of vehicles. There are two parameters within the vehicles which are critical. The first is just how much flexibility can be permitted by the vehicle between the front axle and the following axle. Secondly the speed at which the vehicle can move one axle relative to another.

Within British Rail the critical vehicle for a number of years has been the short 3 m (10 ft) freight wagon travelling at thirty to forty miles an hour.

These vehicles have shown a marked tendency to lift off track where the twist is 1 in 120, i.e. a twist of 25 mm, further other factors have also led to wagons becoming de-railed at 1 in 200, i.e. 15 mm twist. Therefore the minimum standard specified on British Rail, indeed the immediate action trigger is a 15 mm twist measured over 3 m (10 ft) or a 25 mm twist measured over 5 m (15 ft).

It is important to say at this stage that you cannot identify twist by simply looking at the track and further measurement of the track can be deceptive, as voids in track can often give the impression that the twist is not as severe as it in fact is.

The simplest way of measuring twist is to use a cross level gauge on every sleeper and then see what the difference is over four sleepers. However, again one needs to take into account the effect of voids which can make this situation more dangerous.

The actual parameters used for twist on British Rail for three metre are design minimum 7 mm, desirable minimum 10 mm, absolute minimum immediate action correction 15 mm.

For 5 m twists the parameters are design 12 mm, desirable 17 mm, absolute minimum immediate action correction 25 mm.

Dynamic cross level

Ideally the dynamic cross level variation from norm should be zero and that is the basis on which we design.

The desired tolerance is less than 6 mm and if you have a 12 mm variation this requires immediate action. It will be appreciated that there is a close relationship between dynamic cross level and twists although if the dynamic cross variations are very short this produces a bumpy ride rather than a twist derailment type ride.

Top

On British Rail we measure four parameters. On the left rail we measure tolerance on top above and below datum and on the right rail similarly. This would be the parameter that among other things measures dipped joints. The figures that British Rail use have come from empirical measurements taken in the past. Clearly the design is for zero deviation for top. The acceptable is 5 mm above datum or 9 mm below. The minimum immediate action figures are 8 mm above or 10 mm below.

It will be seen from this that the really crucial issue is the minimum action for below datum, mainly dipped joints, where if the dip is more than 10 mm you have serious problems with breakages, cracking and fatigue faults to the fishplates and rail ends.

Alignment

It is clearly desirable to have minimum deviation from a designed align-ment. However, up to 8 mm is allowable but remedial action must be taken when this figure is exceeded. A deviation of 9 mm from the design over a short length is likely to put unacceptable forces into the rail and result in rapid deterioration.

Gauge of the track

Clearly again the ideal figures here are zero. However, it is considered desirable, mainly because of manufacturing tolerances on components, that gauge should not vary more than 3 mm. The minimum acceptable on British Railways and the trigger at which immediate action is required is 10mm. Experience has shown us that if 10 mm has been exceeded then the fastenings or the sleepers are probably beginning to deteriorate to an unacceptable level and if this is not corrected the track will continue to spread wide to gauge resulting in the derailment of the train (see Table 1).

Table 1. Some BR track tolerances (in millimetres)

Parameter	Design	Desired	Minimum
Twist 3 m	7	10	15
Twist 5 m	12	17	25
Dynamic cross level	0	6	12
Top left above	0	5	8
Top left below	0	9	10
Top right above	0	5	8
Top right below	0	9	10
Alignment	0	8	9
Gauge	0	3	10

The wider aspect

The first part of this Paper has considered the nuts and bolts of good railway track engineering. However, it is generally true to say that most tolerances have derived from safety considerations with higher 'comfort' considerations involving tightening of them. This second part looks at more general matters and makes some suggestions as to their relationship. Fig. 1 gives an overview.

Minimum standards

Track and rolling stock

As considered earlier, permanent way engineers maintain track so that trains do not come off. Mechanical engineers maintain stock so they do not fall off the tracks. It is not clear to the Author what came first!

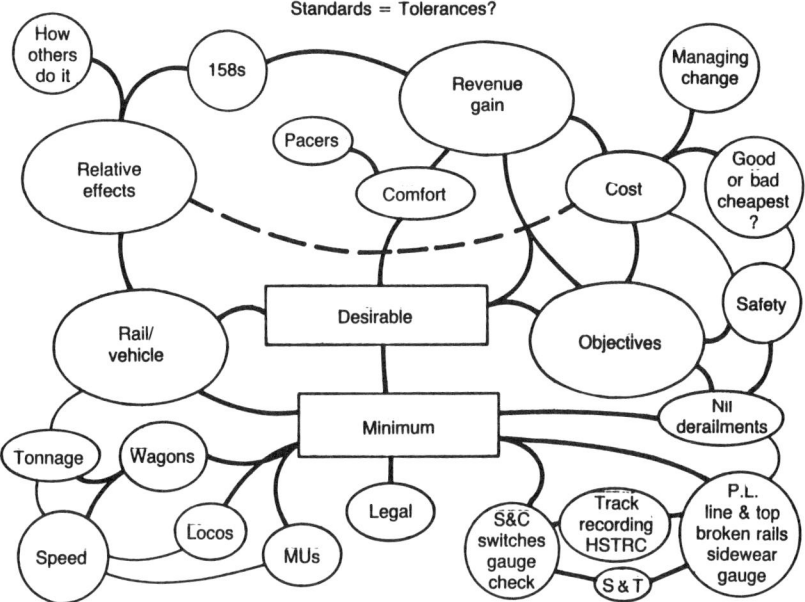

Standards = Tolerances?

Fig. 1. Relationship between standards and tolerances

Legal

In all of the above the legal aspects are not far behind. British Rail is policed by a Railway Inspectorate and after major accidents like Clapham (Southern Region), formal inquiries conducted by Queen's Counsel often follow. Failure to make the right engineering or organisational decisions, or more specifically failing to decide upon the right tolerances and ensuring they are adhered to, will bring ever increasing retribution.

Desirable standards

Objectives

The defining of objectives for permanent way engineers is not as easy as it first looks, once we move above minimum standards.

Safety

Having kept trains on the track, safety considerations dictate that movement of vehicles is not too violent lest it harms passengers or staff on board.

Cost

Cost of maintaining any standard needs to be understood. A simple search for ever higher standards after they have met an objective is counterproductive. It does not reflect well on engineers and drives the railway business into an ever more precarious financial position.

Is good or bad cheapest ?

The assumption that poorer track is cheaper to maintain than good track also needs careful consideration. Very good track on a light axle route needs very little maintenance to keep it in condition. Very bad track on heavy-axled freight lines needs constant attention. The true position is not clear, but it is affected by many variables, not least the quality of material used in renewal. It is the Author's view that many permanent way engineers are maintaining track, with a historically stable traffic pattern, at a cost effective level. This level is often above minimum safety standard and provides an improved comfort level purely by chance.

Managing change

Cost control implies change and change in terms of railway track needs to be handled very carefully. The life of track can be greater than fifty years and the formation considerably longer. A policy change anywhere in the cycle can start a trend which may not easily be reversed. For example, when considering a route maintained by sleeper replacement, if one decides to leave wooden sleepers in the track for a longer period, the further deterioration of older sleepers may bring about early damage to new sleepers.

On the other hand, on routes where the policy is to relay, replacement with track to too high a specification produces no problems for the engineer, other than driving his organisation into long-term financial crisis.

The upshot of all this is that changes should be carefully monitored and the engineer should have a clear grasp of the tolerances he is working to and why. This is particularly true where traffic patterns have changed.

Rail/vehicle

Whilst minimum standards for both civil engineers and mechanical engineers must be (and are) agreed in order to achieve nil derailments the relationship between desirable standards is much less clear. The classic question has always been 'what type of steel should form the wheels and what type of steel should form the rail'. Wear rates will vary greatly on wheels or track dependent on the solution chosen.

Minimum standards will then dictate the maintenance frequencies where safety is an issue, but an imbalance between the two will cause undue high frequency maintenance cycles for one. Unilateral action by one group to improve wear characteristics can have a dramatic down side for the other.

It therefore follows that standards should be developed by mechanical and civil engineers jointly.

British Rail has developed complex computer tools to analyse the behaviour of track for given vehicles that are proving useful in predicting the effect of change.

Relative effects

The ideal position is to achieve a set of tolerances that meet the comfort requirement of the total traffic flow whilst minimising the expenditure of both the civil and mechanical engineers.

It is worth noting that this ideal is theoretically possible to achieve for railway engineers, whilst it is not open to highway engineers.

The problem should be simpler where there is only one basic type of traffic flow. A recent British Rail example is the Settle and Carlisle Line that was under threat of closure for some years. Tailoring the track maintenance to a now mainly 'Sprinter' train service has enabled civil engineers to improve ride quality whilst halving track maintenance costs.

How others do it

Given the diversity of railway networks throughout the world, it is reasonable to assume that most combinations of track/vehicle relationships have been tried. The dissemination of lessons learned by individual networks probably holds the key to optimum solutions.

Comfort

Comfort is largely an issue for passenger railways. Surveys on British Rail suggest that customers do not rate ride comfort very highly on their list of priorities. This may be a result of generally high standards of ride. Comfort becomes an issue only where the most cost-effective solutions produces an uncomfortable ride that begins to deter passengers from using the railway.

Regional Railways stock

Regional Railways policy has been to replace stock with light axled Multiple Units. Some have been more successful than others. The "Pacer" that was based on a bus chassis has developed on unfortunate reputation with engineers because it has a tendency to hammer down joints. The ride is also poor with customers referring to them as 'Nodding Donkeys'. Sprinters and even more so Class 158s are a new generation of light axled multiple units bringing a high quality product to the customer whilst decreasing wear and tear to track. It is a good combination which suits rural routes.

Revenue gain

The values of minimum tolerances should be those that result in nil derailments. This is the lower limit to which railway engineers can work. The upper limit must be perfection. The desirable tolerances are those that ultimately produce the greatest Net Revenue Gain. The Author does not specify what those are. Because of the issues raised in this Paper, they are infinitely variable. Each railway and indeed each route must be analysed to determine the appropriate values.

4. Upgradation of track for higher speed and rectification of long wave length defects in track geometry

M. SESHAGIRI RAO, FICE, Rail India Technical & Economic Services (RITES)

Background

With intense competition from road and air, it is becoming increasingly necessary for railways to upgrade their tracks for faster passenger and heavier freight trains. In India, tracks originally built 140 years ago for 75 km/h and 15 tonne axle loads have been upgraded up to 140 km/h and 23 tonnes by improving the standards and geometry of existing tracks. While axle loads may not be increased further due to the restrictions imposed by bridges, the train loads will keep rising. 4500–5200 tonne trains are now commonplace and 9000 tonne trains are running on nominated routes.

Introduction of long welded 60 kg rails, prestressed concrete sleepers with elastic rail clips and a ballast cushion of 25–30 cm has helped obtain a track structure fit to cope with high speeds and densities of 50–60 gross million tonnes a year. Rails are invariably of high manganese wear-resisting quality. Long welded rails are the rule and welding is done by the flash butt process. Monobloc PRC sleepers are being extended to turnouts and perhaps even to girder bridges in future.

Modern traffic imposes a lot of punishment upon the permanent way and a continuous monitoring of track condition is essential. Rail fractures are forewarned by ultrasonic rail flaw detection. Track irregularities are monitored by Electronic Track Recording cars and the vehicle reactions by oscillograph cars.

Rail/vehicle compatibility

With the gradual exit of steam traction, of laminated springs, of screw couplers and of four wheeler wagons, the basic philosophy in the measurement and rectification of track defects has necessarily to undergo many changes. It is not enough to merely have a better track structure.

Vehicles have changed in a big way. Springs of high speed coaches are now very soft and well damped whether they be of coil steel, rubber or air. They have comparatively longer periods of oscillation and greater amplitudes. Three-piece bogies for eight-wheeler freight wagons and automatic centre buffer couplers also have certain rigidities which have introduced certain hitherto unknown effects on track.

Laminated springs of goods wagons oscillated at 0.2-0.4 sec (2.5 to 5 Hz) and had a limiting amplitude tolerance of 10-12 mm. They were highly sensitive to twist on short chords but not to unevenness. Coil springs of modern wagons oscillate at 0.25-0.8 secs (1.25 to 4 Hz) and have a free amplitude of 20-35 mm. ICF high speed passenger coaches have a period of 1.2 to 1.4 sec (0.7-0.8 Hz) and tolerance of 45-50 mm in amplitude.

Increase in oscillation periods has increased the wavelength for track parameters almost tenfold even while the tolerance to amplitude has improved. Thus, while twist is easily absorbed by modern stock, unevenness over long chords has taken its place as the critical parameter for determining the accelerations. Yet the older vehicles also continue to run and track maintenance can not disregard the old norms. Unlike yesteryear, we now have to measure the track defects in various forms characterised by their wavelengths which may be split into three groups.

Chord lengths for track measurements

Firstly, we have short wavelengths of 50 mm to 2 m. These would be the defects associated with the rail shape, such as hogged ends, alignment kinks, corrugations, imprecise welds, loose packing and the junctions between hard and elastic beds. They need correction by grinding, de-kinking and firm packing.

Medium wavelength faults between 2 m to 25 m would be those caused by degradation of ballast, settlements caused by rains and trains, looseness of fittings, etc. They can be corrected by hard and firm packing resulting in an improved track modulus. These are easily eliminated by the tie tamper working in the automatic mode, by off-track tampers, by measured shovel packing, or by any traditional method.

Long wave faults between 25 m and 125 m might often be due to settlement of embankments and other long-term ground movements. However, many more of them usually exist by design (or the lack of it) particularly over routes where economy in earthwork was the supreme need during the original construction. These are most critical to upgradation of speeds, for the safety of long and heavy freight trains and for fourwheeler centre buffer stock running light.

Removal of such defects is a fundamental prerequisite to the introduction of higher speeds. A vehicle running at 140 km/h and vibrating in the vertical mode at 0.8 Hz would need an unevenness on an 80 m chord of generally no more than 20 mm and exceptionally 30 mm. At the other end, the leaf sprung four wheelers impose a limit of 5 mm on a 10 m chord.

Track geometry spectra viewed through Power Spectral Density functions show how variance of the energy of track irregularities get distributed over the wavelengths. The rate of deterioration of track is very different between two sections even if the traffic patterns are the same. Deterioration rate varies exponentially with the faults and the overall effects keep extending to longer and longer wavelengths if maintenance is disregarded for long periods (ref. 1).

Traditional correction systems

Certain track parameters are measurable in absolute terms e.g. gauge, cross level, sags and creep. Certain others can only be measured in terms of single or double differentials, e.g. twist, alignment and unevenness. The directly measurable defects were, till about thirty years ago, the only items attended to during track maintenance in the developing countries.

Traditional track maintenance involved adjustment of gauge and cross levels, slewing to alignment and packing ballast under the sleepers to give a firm consolidated bed to give a uniformly sag-free longitudinal plane to the rail top.

With modern track machines, the tamping and levelling is a combined activity wherein the rail top profile is designed automatically as the machine moves along. A three-point system using an infra-red beam defines the design. The two extreme points are located on the packed track and old track respectively. The tamping tool at the middle third point lifts the rails to a level so determined as to give a straight line between the other two (Fig. 1). Thus the automatic tamping and levelling system is theoretically able, in one pass, to effectively improve unevenness over chords up to 10 or 12 m. Tamping machines are affected by certain limitations imposed by rail stiffness, track settlement under the weight of the machine and inadequacies of ballast support which reduce this effective chord length further.

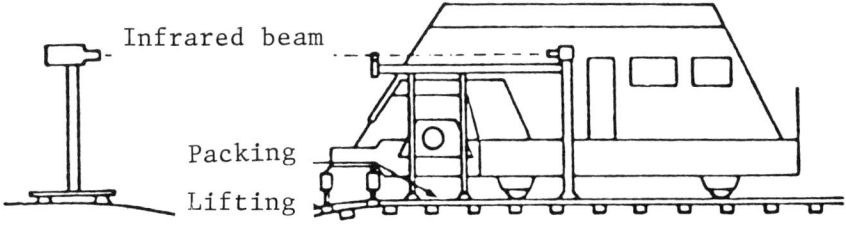

Fig. 1. The tamper principle

The manual systems also follow practically the same principle whether by measured shovel packing, by off-track tampers, or by beaters. Unevenness corrections are restricted at the most to chord lengths of 8-10 m.

Whatever the system adopted, the correction lifts are generally small due to the very nature of the short chord lengths. Such small lifts generally result in rapid settlements and the effects do not last long. To obviate this, it is usual to apply an extra general lift of around 25 mm. It has therefore now been accepted that traditional methods of track improvement either using automatic machines or by other means are limited only to short chords and have little effect on long chord irregularities (ref. 2).

Theoretically speaking, repeated passes of an ordinary tie tamper could improve long chord defects too. Such a process will result in very large lifts resulting in wastage of ballast as well as of the occupation time.

This has been improved upon by the new laser-guided equipment which permits the use of very long reference beams and lifting to the line defined by the laser. This system too is subjective and can require excessive quantities of ballast and extra passes may be required if the desired long chord unevennesses are not achieved in a single pass.

Curve re-alignment

As is well known, the effect of lateral defects on safety at high speeds is very much more severe. Tolerance to vertical defects is dependent up to the spring amplitude which can be improved by vehicle design. Tolerance to alignment defects is limited by the play of the wheel set between the rails which is severely restricted particularly in high speed vehicles.

Ride comfort too is dependent very much more upon lateral track defects than on vertical ones and is dominated by low frequency oscillations caused by long wave track imperfections in the horizontal mode. In general an amplitude of 5-10 mm on an 80 m chord length is the limit beyond which a lurch may be expected.

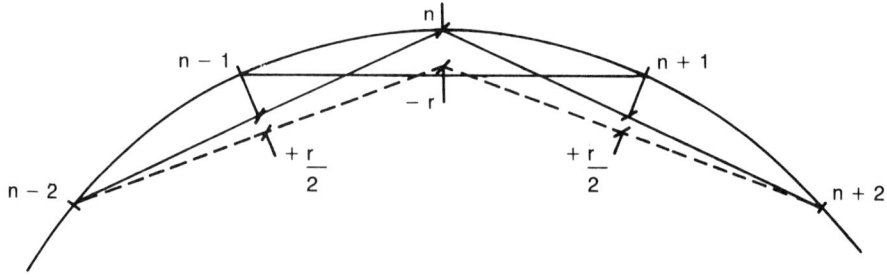

Fig. 2. Example of automatic lining system

Various types of automatic lining systems for improving alignment exist. Most common of them (Fig. 2) measure the track over a short length and equalise the curvature between 3 or 4 points on that length (refs 3 and 4).

In the more modern versions (ref. 5) the machine measures the track geometry over the entire curve on the first run and an on-board computer carries out the design of the appropriate slews for the desired versines based on the Hallade technique. It does not however ensure perfect trapezoidal versine diagrams and does not permit transition shapes to one's specification.

Simple smoothing, whether by machine or by manual, means does not improve long chord track faults except after a very large number of passes. In that case the lifting and slewing efforts are many times what an efficient design can give.

The Top Track system

Whenever there is a need to upgrade track geometry, it is therefore necessary to overcome these pitfalls through a design in totality and to apply the corrections in the design mode as opposed to the automatic mode.

The system, christened Top Track, follows the sequences as shown in the flow chart (Fig. 3). 10 m points (called 'stations') are numbered on the web of the rail to define the location points and benchmarks fixed every 250 m. The vertical alignment of the existing track is measured to an accuracy of 1 mm using precision levels. Intermediate levels are obtained at salient locations.

Curves are measured over a 20 m chord which is the universal practice. If there is heavy wear on the gauge face of the outer wheel, the versines can be measured either over the inner rail or optically along the centre line of the track.

The design of the proposed track profile is carried out in the office by an off line computer. It is therefore necessary to record the limitations of lift and slew at each salient point for making sure that the design does not infringe the fixed installations and structures.

The heart of the Top Track package is its highly sophisticated design software.

Design features

In respect of laterals, it normally gives a perfectly trapezoidal versine diagram with a guaranteed lowest slewing effort and free from point-to-point versine variations. The slews so obtained are so small that fixed structures rarely come in the way. By and large, the transition lengths obtained by this method are also more than adequate for the specified requirements of cant gradient and cant deficiency gradient of the railway

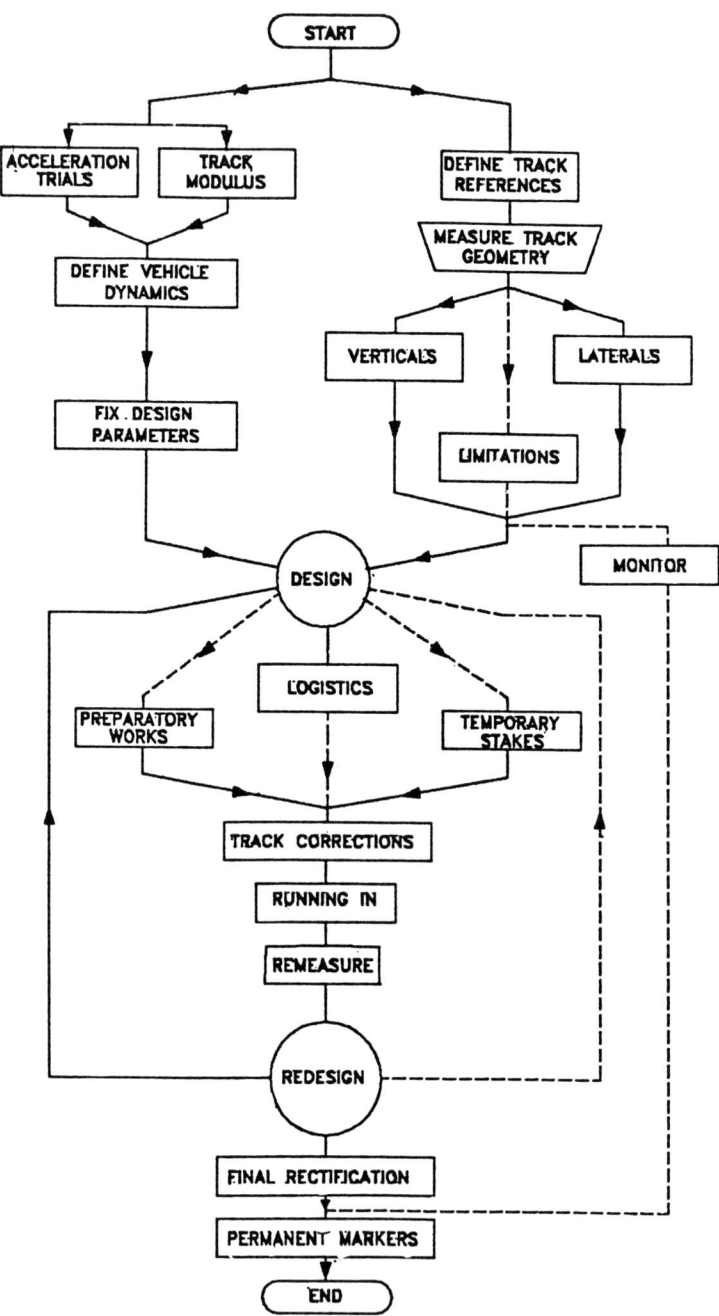

Fig. 3. Top Track activities

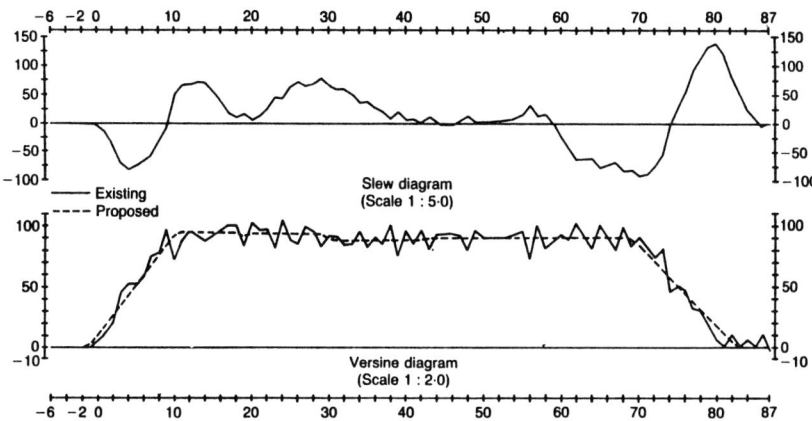

Fig. 4. W. Railway Curve No. 25

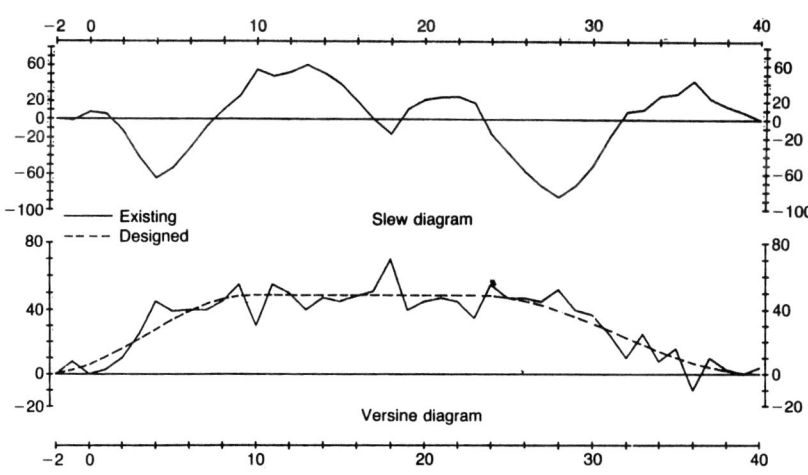

Fig. 5. W. Railway Curve No. 24 sine transition

57

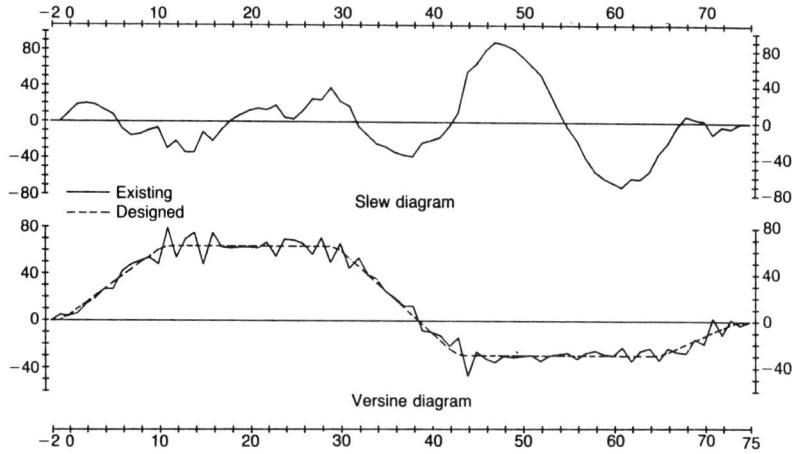

Fig. 6. W. Railway Curve Nos 34 and 35

administration.The programme, however, easily permits provision of any transition lengths of one's choice.

In all cases, the sums of outward and inward slews are equalised. This ensures that there is no elongation or contraction of the long welded rail and the slewing of curve does the not result in induction of high stresses which might destabilise the slewed track (Fig. 4).

As in many other countries, we too have the problem of the rigid 3-piece bogie designs for freight wagons being adopted for its simplicity. They are however very harsh on track and impose very severe lateral forces particularly over transitions due to high torsional moments.

It has been found that use of sinusoidal transitions instead of the usual cubic parabola helps reduce these lateral forces (ref. 6), the Top Track computer programme can provide transitions in the form of cubic parabola, sine. The curve, cosine curve, Bloss or Double-S (ref. 7) equalisation of outward and inward slews is ensured in all the cases (Fig. 5).

Experience has shown that a truly circular curve never goes out of alignment and retains its shape for very long periods needing maintenance, if any, only over the transitions where the curvature is varied. Even this drops substantially with sine transition curves.

Notwithstanding this, there could be a few locations which might still face the problem of slewing space being restricted to tighter limits than the least calculated slews for such reasons as closely founded traction masts and closer track centres. In such cases, it is possible to apply a correcting contra slew which is evened out over a long length causing very minimal distortion to the perfectness of the trapezium.

We have so far dealt only with normal circular curves having transitions at the two ends. Top Track also has programmes for compound curves and reverse curves ensuring minimum slewing (Fig. 6) effort and perfectly trapezoidal versine diagrams.

In the case of reverse curves between parallel tracks however it is more desirable to provide a sinusoidal alignment particularly if it is undesirable to have the intervening straight.

Vertical profile

In the field of vertical alignment, the programme ensures that the lifting effort is the absolute minimum required for arriving at the chosen limit of unevenness. The aim is to ensure an unevenness limit of 20 mm on 80 m chord, 10 mm on a 60 metre chord and below 5 mm for 40 metre chord and below. It has been found more economical to design in such a way that about 90% of the total length is lifted and about 10% lowered in the process. This keeps the ballast consumption down to an optimum minimum (Fig. 7).

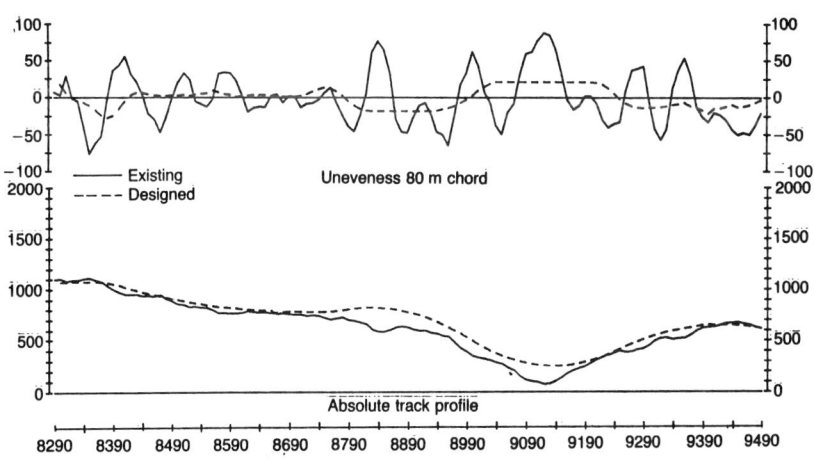

Fig. 7. W Railway Godhra Ratlam section

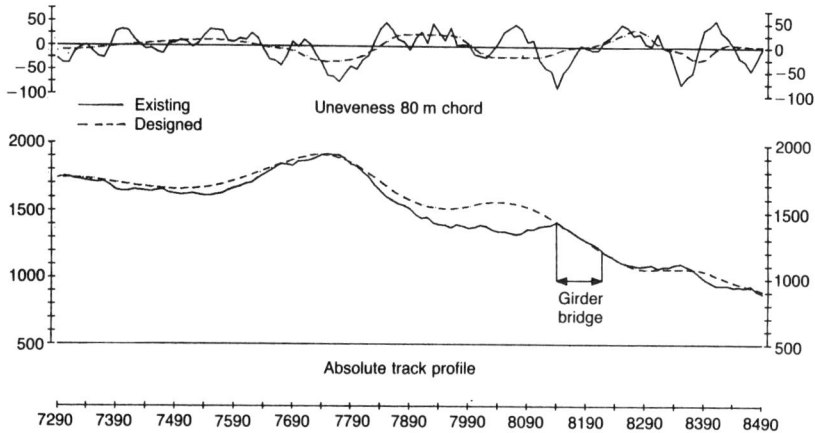

Fig. 8. W. Railway Godhra Ratlam section

However, it has to be appreciated that the vertical profile is severely restricted by a large number of limitations. While taking the levels of the rail top, levels are also simultaneously taken of the hard formation under the sleepers in order to get the depth of the existing ballast cushion. In stretches where the existing cushion is inadequate, provision is made in the design for lifting the track to such a level as to provide for the minimum ballast cushion requirements as well.

While the horizontal curves are by and large realigned to their original construction location and are therefore not subject to severe constraints, the vertical profile has usually undergone settlements during the course of its life and also been lifted during several stages of new ballasting. It is therefore subject to larger corrections than the laterals. Yet, girder bridges can neither be lifted nor lowered. Flyovers, road bridges and foot bridges can only be lifted or lowered up to a very limited extent. Certain other locations such as tunnels and cuttings in hard rock cannot be lowered at all. Points and crossings, passenger platforms and overhead electric traction wires, all impose restrictions on the amount of lift (Fig. 8).

While every effort is made to get a design with the prescribed unevenness, small compromises do become unavoidable in certain difficult locations, particularly adjacent to girder bridges. We have so far not had to compromise beyond 28 mm on 80 m chords, i.e. 7 mm on 40 m chords with no compromise at all for shorter chords.

In the accompanying diagrams, the results achievable through Top Track technique vis-a-vis machines running in the automatic mode are clearly explained.

Field execution

Top Track can be executed mechanically, semi-mechanically or manually. However, machines have their own limitations. The modern Plassermatic machines can give single lifts up to 75 mm or so and slews up to about 50 mm. In case design demands more than this, it is necessary to resort to other methods.

The most common method is to use hydraulic jacks in both horizontal and vertical modes after loosening the ballast from the cribs and the shoulders. This work is done under speed restrictions. It is also necessary to readjust the overhead traction wires if shift in the alignment is substantial.

Permanent markers are fixed along side the track every 10 m to define the design level and lateral position. The initial rough lift and slew are provided by hydraulic jack as stated above and position is fixed within 5 mm of the design to the best of the ability of the equipment. Thereafter the final packing, levelling and slewing is carried out by the tie tamping machines of whatever model available.

Thus in reality the Top Track system merely implements the well known principles for upgradation of track for high speed, namely that the variations of track defects on long chord measurements in the region of 60-80 m have to be kept down well below the spring amplitude. While automatic machines may appear to be a very easy and simple way out, they are not in reality able to satisfy the long chord parameters. Since most of our railway lines were not constructed with such yardsticks in view at all, an extremely delicate design is required to achieve such parameters with minimum effort. This is possible by working from whole to part and not from part to whole.

Conclusion

The results so far achieved have been extremely gratifying and lateral accelerations at speeds up to 130 km/h have shown drastic reduction both in lateral and in the vertical mode after Top Track treatment and chronically derailment-prone section have become accident free. The Top Track system of track maintenance design is well on its way to universal acceptability. In the field of laterals alone, a large number of curve realignment problems are being solved on fax-to-fax basis. This is being utilised in five countries around the world and by five railway networks in India.

References

1. TAKAI H. Maintenance of track with longwave irregularity on the Shinkansen. *RTRI* , August, 1990.
2. REISSBERGER K. Key elements in the maintenance of high speed track. *Railway Gazette International., March, 1989.*
3. ESVELD C. *Modern railway track.* p 223.
4. MUNDREY J. S. *Railway track engineer.* Tata McGrawHill, p. 276.
5. JOHNSON D. M. Laser guided tampers improve track geometry. *Railway Track & Structures.* April, 1991.
6. MIELOSZYK E. and KOC W. General dynamic method for determining transition curve equations. *Rail International.* October, 1991.
7. GUBAR J. Railway transition curve methods. *Rail International.* April, 1990.
8. GOEL R. K. Top Track smooths out derailment problems. *Railway Gazette International.* April, 1992.

5. Inspection, monitoring and measurement of track condition

D. R. BALLINGER , Transmark, UK

Background

The intention of this Paper is to address the subject in road terms but with particular emphasis on the practices suitable for adoption on a low speed (80 km/h), low tonnage (2 million gross tonnes), and low budget railway. The Author has drawn on his recent experience in Botswana in central southern Africa, and neighbouring countries on the 3ft 6 in. "Cape" gauge system. By way of comparison, there are some references to past and current practices on British Railways.

Introduction

The need to maintain adequate standards of track for the traffic being carried is a fundamental principle of railway permanent way engineering, the basic objective being to avoid derailments and provide acceptable ride conditions at no affordable cost. Whilst it is a practical impossibility to prevent derailments completely, in particular those due to causes which are not track related and thereby largely outside the control of the civil engineer, it is possible progressively to reduce the likelihood of incidents by directing the organisations attention to those aspects of track conditions which feature in derailments most frequently, which are twist faults, cyclic top faults, buckles and of course broken rails, along with switch and crossing layouts in general. Priority must of course be given to the elimination of defects on the open line, on the basis that derailments outside stations and yards usually occur at higher speed and are not only more damaging and dangerous, but also more disruptive and difficult to clear up.

In order to check that track conditions are being maintained to adequate standards, a regime of inspection measuring and monitoring will have to be devised and put in place. To do this the engineer will have to take into account many factors, ranging from the expertise of his individual staff and the training, or re-training needs, through to the availability of any mechanical or electronic track recording devices.

An extremely important factor influencing the above will be the cost and cost benefits of the various alternatives being considered.

Sophisticated track measuring vehicles are expensive (from US$2000-5000 per shift plus travel costs), and the justification of this sort of expenditure on a regular basis is difficult to contemplate for most low budget operations.

Conversely, manual inspection is relatively cheap, even if carried out on a daily basis, but the findings tend to be subjective and do not lend themselves to any further analysis to determine whether any trends are being established.

Clearly the ideal monitoring regime will include a mixture of manual inspection and machine measurement, and there are no hard and fast rules as to which is the best combination for any particular set of circumstances. The following paragraphs discuss the options available and the considerations which have to be taken into account.

Inspection

Inspection consists of the visual checking of track condition either on foot or from a rail mounted vehicle. It is a vital and mandatory first level of monitoring and in some situations it will be sufficient in itself for the needs of the administration. It might even be all that the administration can afford, or is prepared to justify, but as stated earlier, it is obviously a subjective monitor of track condition and is generally only relied upon to give an overview of the track conditions prevailing at the time of the inspection. Beyond that it relies on judgements being formed from a comparison of opinions and gut-feelings over a series of inspections on the particular stretch of track in question.

Inspections are usually carried out at the following levels

- Patrolman - 2 or 3 times per week
- Inspector - monthly or bimonthly
- Engineer - 6 monthly or annually.

Patrolman

The Patrolman is concerned mainly with checking the track for defects requiring immediate or urgent maintenance attention. He obviously needs to be sufficiently experienced and trained be able to recognise defects and categorize them in terms of severity. He must be capable of instituting emergency procedures if traffic safety is threatened.

He needs to be able to read and write in order to absorb written instructions and convey his impressions to others via his inspection notebook.

A good Patrolman is an extremely important and often undervalued member of the inspection team, but it is vital that the right man is selected, and that adequate training is given to ensure that he knows what is expected of him. The feedback from the Patrolman can and should form the basic information source for a simple work planning system, and as an indicator to where higher levels of inspection should be directing their attention.

Inspector

The Inspector is the first level of inspection where any subjective monitoring of standards is possible, using gut-feeling and recollection to determine whether the track is getting better or worse. It is also the level where cost of the day to day work planning is initiated. Ideally the inspection should be carried out on foot, but in remote areas some form of self propelled trolley is required. Rudimentary inspection trolleys are usually very poorly sprung and are therefore ideal for reminding the inspector how bad his joints are - a fact which might escape him if he relied solely on foot inspections. An even better appreciation of the true riding quality of the track can of course be gained from the cab of a service train at line speed. Whilst this form of inspection if highly indicative, in terms of highlighting the bad spots, it is no substitute for the detailed scrutiny carried out at ground level.

The Inspector should be concerning himself with the general condition of his track and plan whatever day to day maintenance is required using his own resources. He should also build up medium term planning to include tamping machines, as they are allocated to him, and component renewals in turnouts or on curves.

Formal reports of the inspections with detailed commentary on the findings should be submitted to the engineer regularly.

Engineer

The Engineer is primarily concerned with ensuring that standards are being adequately maintained and that resources are being used effectively. In addition, he should be formulating longer term strategies for the track, covering both maintenance and renewals.

It is normal for this level of inspection to be carried out by a motorized inspection trolley, with short stretches being covered on foot as necessary. It is advantageous and desirable for the Engineer to be accompanied by his section Inspector so that problems and strategies can be discussed first hand.

Such inspections present an ideal opportunity to meet the workforce in their working environment to discuss the work being done and to check that it is being done properly. It is also a useful forum for taking on board any domestic problems or grievances held by individuals or groups. All major structures should be given a superficial inspection with particular attention being directed towards looking for evidence of scouring or embankment erosion. Where track is carried on bridge timbers, directly fastened to the bridge girders, these should be checked for condition, security of fastenings and squareness and regularity of spacing.

Inspection (examination) of rails

As well as the inspection of the geometric condition of the track it is vitally important to inspect the components for wear or failure. By far the most important component of any railway is, of course, the rails. A few fastenings missing here and there, or the odd broken or rotten sleeper will not normally cause any problems provided the situation is kept under control. Rails of course, are different, because as all permanent way engineers know, it needs only a few cm of rail to break away and a disaster is waiting to happen.

Broken rails are fairly normal occurrences on most if not all railways, and clean breaks away from the rail end are usually easily discovered by routine inspection and rarely lead to further problems. Broken rails at fishplated joints are potentially very dangerous, and also not very easy to detect by normal visual inspection.

Broken fishplates are sometimes regarded as a trivial almost everyday occurrences, but they can be an indicator of more deep-seated problems. Joints which suffer broken fishplates on a regular basis always have other fundamental defects such as voids, fouled ballast, formation problems and crippled rail ends. The inspection and monitoring process should draw attention to such joints as requiring detailed examination. It is almost inevitable that fatigue cracks will be developing at the bolt holes, and failure to detect these early could result in a catastrophic collapse of the rail end the next time the fishplates break. Historically on British railways, rail ends were cleaned and visually examined annually when the joints were stripped for re-oiling. If cracks were suspected, then dye-pentrant was used to confirm the nature and extent of the crack.

Visual inspection could never be relied upon to pick up all cracks and was absolutely useless for detecting internal defects in the rail, such as tache ovales or piping, both of which might eventually lead to cracking and rail breakages. Hand propelled ultrasonic testing equipment with associated angle problems will detect almost all internal defects within the rail, and confirm the existence, nature and length of all developed cracks including those in the vital rail-end zone.

Inspection of rail ends should be carried out at least annually, by ultrasonic means wherever possible. If such equipment is not available, visual examination by trained staff under strict supervision, is infinitely better than nothing but takes considerably longer. Ultrasonic equipment is not expensive at about US$4000 per unit, especially when compared to the horrendous cost of dealing with a single major derailment due to a broken rail end.

Measurement

Whilst it is possible to carry out spot measurements of track condition parameters, such as gauge, cross-level, twist etc., manually, it is not practical to get any overall impression of track quality from these, nor is it possible to compare readings on subsequent inspections to determine quality trends.

What is needed is some form of mechanical measurement coupled with a digital print-out combining the important parameters in a track quality index which can be used to compare with previous and subsequent runs.

British Railways relied on Matisa trolleys for many years or track geometry recording on all lines. This apparatus used physical contact by wheels and feelers transmitted to the recording pens by chains and pulleys and was thus prone to inaccuracies caused by wear and/or setting problems. In its original form, the track parameters were plotted on a continuous analogue trace, with exceedencies for twist over 3 metre base being recorded on the trace and marked on the track by red or yellow paint splashes.

The usefulness of all the above information was however limited by the fact that it did not give any other than a visual indication of track quality.

In order to address this deficiency and provide a quick and comparable reading of track quality, British Railways and North Eastern Region jointly developed the NEPTUNE system.

NEPTUNE was credited as being 'a complete system or track fault analysis and correction, providing for the planning of track maintenance, the efficient use of manpower and on-track machines, the forecasting of track deterioration trends, cost control and plant requirements' (ref. 1).

To varying extents, the system was capable of being all of these, depending on the recipient end-user, and it was eventually adopted almost universally throughout BR. Various other devices were also in use such as hallade cars and reidmeters, but none of them achieved the universal acceptance of the Matisa with Neptune.

Track measurement for high speed running

For speeds of 100 miles/h and above, a very high quality track must be provided and maintained. It had long been realised that measuring the quality of such track by a lightweight, low speed recording vehicle was not a satisfactory situation, even if the accuracy of the machinery could be relied upon, which it could not. The late 1970s saw the introduction of the High Speed Track Recording Coach (HSTRC), which employs a contactless system of light beams and cameras incorporating transducers and accelerometers to measure the geometric parameters at full line speed. In addition to the 10 line analogue trace, (an example of which is shown at Appendix 1), track parameter exceedences are continuously measured and summed as standard deviations for each $^1/_8$ th mile section (200 m) of track (see Appendix IA).

The acceptability and reliability of the system coupled with continued improvements and developments have seen the HSTRC and its sister derivatives, adopted for all running line track recording on BR.

Such a sophisticated piece of equipment is very costly, not only to develop, but also to maintain and operate. Its use can only be justified on cost grounds for high speed, high tonnage lines, although it is used extensively on BR's secondary lines by virtue of being the only equipment currently available.

Many of the major railway administrations in the world have their own in-house developed track recorder, such as the Netherlands Railways (NS) 'BMS' system (ref. 2), but for smaller administrations, such prohibitive development costs would be unjustifiable.

There are though, a number of commercially available machines on the market for purchase or hire with varying levels of sophistication in terms of equipment and output information. It is therefore possible for even the smallest railway administration to monitor its track standards effectively through some form of regular track recording.

Plasser and Theurer manufacture a number of track recording vehicles starting with the EM 30, which is similar in some respects to the Matisa trolleys formerly used on British Rail, through to the highly sophisticated EM 160 which uses a high speed passenger coach as its base vehicle. There is also a road-rail version, known as the EM 25 which could clearly be very useful where lines are not connected to major systems and thus afford no access for conventional on-track recorders.

In common with most other modern track recording vehicles, the parameters recorded by the whole range of these machines include

- twist
- versine (left rail and right rail)
- top or rail surface level (left rail and right rail)
- gauge
- cant or superelevation
- measuring vehicle speed
- mileage or kilometre marking
- feature marking (stations, junctions bridges etc).

The first five of which correspond to 'the 7 classic track parameters' (ref. 3), with the remaining three being for general orientation.

Appendix II shows a section of a typical Plasser EM 80 analogue trace of all the aforementioned parameters, taken from an actual run on one of the routes in Botswana. In addition to the recording, marking and audible warning of serious exceedances, in twist and occasionally alignment, on board computers record the collected information digitally, and it is from this data that the statistical and comparative analyses of the state of the track are derived.

Appendix IIA shows typical printouts from the the same EM 80 run listing exceedences in top, twist and alignments, along with derived maintenance planning data.

In all track recording and measuring systems, exceedence levels are set at appropriate levels for the class of line being recorded, and it is these levels which form the basis if the derived analytical output used in maintenance planning and monitoring.

The range of off-line output available varies between systems, but can usually be tailored to suit the organisations specific requirements, at little or no extra over the cost of carrying out the recording.

Track/vehicle interaction measurement

The way a rail vehicle reacts when travelling over a stretch of track is a fair indicator of the state of the track in terms of its geometric quality under load. Instruments to measure these reactions, such as the Ridemeter (ref. 1 - p. 199), have been around for many years to measure and monitor, through comparison, the quality of ride experienced by passenger coaches. Any measurements taken were of course always subject to the vagaries of the vehicle characteristics, and readings taken over the same stretch of track, but on different vehicles, could never be compared with any degree of reliability. It nevertheless gave a useful indication of the ride quality and picked up any serious lurches enabling early investigation to take place.

Continuing the development of the same principles, Geismar have brought out a family of instruments which they describe as 'a new approach to identification of track and moving stock performance by accelerometry'.

The 'Macminder' is claimed to monitor the performance of both vehicle and track under service conditions. Measurements are made with light-weight, portable measuring units which are positioned on various vehicles in the train. Data is collected throughout the day without the need for staff in attendance and the output is presented in graphical form (Appendix III). In essence, poor vehicle ride experienced by a single vehicle in the train indicates a defect in the vehicle, whereas poor vehicle ride experienced by all vehicles in the train indicates a track condition requiring investigation.

'Macmonitor' is an instrument which is used singly by an operator and is intended to determine the cause of a particular vehicle problem, or to monitor the performance of a particular vehicle either under test or in service. The 'Macmeter' is an up to date version of the ridemeter and is used to measure passenger ride comfort or cant deficiency.

'Macktrack' claims to be portable self contained railway track monitoring system, and perhaps this unit will be of most interest to the permanent way engineer.

Being only a fairly recent development, it remains to be seen how the system will be accepted. Obviously the capital outlay to purchase the equipment will be considerably less than that required for a track recording machine, and whilst it will not produce an intrinsic measure of track geometry, it should indicate the worst fault locations and provide some measure of track quality which can be compared.

Monitoring

As stated above, effective monitoring of track standards is only practicable when a system of regular measurement exists, which will allow comparisons to be made with previously gathered data. If trends are to be discerned reliably, then the information upon which the monitoring is to be based needs to be collected regularly and accurately. Opinions formed from the monitoring of track recorded data should be verified by site inspection. Falling standards as perceived from regular recordings could be due to deteriorating condition, or simply lack of effective maintenance.

Careful monitoring will enable judgements to be made as to where maintenance resources appear to be most needed, but more importantly, it will assist in ensuring that resource allocation is directed to those locations where it will produce the greatest benefit. With experience, wastage of maintenance effort and resources could be significantly reduced.

Directing tamping machines selectively to the worst sections can produce the largest general improvement in the overall standards of the track, assuming of course that the track is fit to tamp in the first place. Tamping of track which is in poor condition will more than likely make matters worse.

On low budget routes, tamping of joints alone is probably all that is needed on alternate tamping cycles. Joint tamping allows the tamping machine to proceed up to four times faster, enabling four times as much track to be treated per shift, with all the cost benefits which that implies.

Conclusion

Inspection, measuring and monitoring are essential processes in the cost-effective management of track infrastructure. Properly organised systems should ensure that adequate standards are maintained through effective resource planning utilisation.

Medium and longer term strategies can be formulated and included in budget forecasts. Track life can be significantly enhanced by monitoring its performance and condition, and tailoring the resource input accordingly. The track structure is the most fundamental and important physical asset of any railway undertaking, but it is often the least regarded and most neglected.

References

1. HEELER C.L. (ed.) *British Railway Track*. Permanent Way Institution, 1979
2. ESVELD C. *Modern Railway Track*. MRT Production, West Germany, 1989
3. RIESBERGER K. Track recording cars - the principle instrument for track maintenance. *Railway Technical Review*. Edition 28, 1986/87, pp 41-46.

Appendix I

```
DIRECTOR OF CIVIL ENGINEERING   TRACK RECORDING SYSTEM (TRC)
STANDARD DEVIATIONS AND LEVEL ONE EXCEEDENCES *** ON BOARD REPORT ***        RUN DATE :080592    [V6.R02(?)]
JOURNEY:                                                   0098/1
REGION:                          AREA:                     PWME


PARAMS   TWCM      MT35      MC35X    AL35     A35X     MT70     AL70     GAUG     ACT LNE  avs   fix
                                                                                  SPD SPD

+ABOVE   10MM      8MM                12MM
-BELOW   10MM      15MM               12MM


MILES    SD  EX   SD  EX+ EX-   sd    SD  EX   sd    SD       SD     MIN MAX MPH MPH

 2/7 stc! 33  2 ! 30   2  0 !   25  ! 66   3 !  62  !  28   !  69  !  31  44!  85 60   410 !

 3/0   ! 42   3 ! 35   2  0 !       ! 71   1 !      !  31   !  77  !  29  48! 105 60
 3/1 stc! 17  0 ! 38   5  0 !   34  ! 34   1 !  34  !  43   !  51  !  33  47!  14 60
 3/2   ! 29   0 ! 28   2  0 !       ! 40   0 !      !  37   !  **  !  31  44!  18 60
 3/3 stc! 24  0 ! 35   1  2 !   31  ! 31   0 !  27  !  41   !  87  !  28  52!  24 60
 3/4   ! 12   0 ! 41   5  0 !       ! 12   0 !      !  53   !  34  !  31  39!  33 60
         225       295             265     362      351       333    677



DIRECTOR OF CIVIL ENGINEERING   TRACK RECORDING SYSTEM (TRC)
LEVEL TWO EXCEEDENCES                       *** ON BOARD REPORT ***          RUN DATE :080592    [V6.R02(?)]
JOURNEY:                                              0098/1
REGION:                          AREA:                     PWME


PARAM        : TWCM   : TWCM   :   LTOP        :   RTOP       :AL35  :GAUG  :L CYCLIC:R CYCLIC:LNE SPD
             :        :        : ABOVE: BELOW: ABOVE: BELOW:      :      : CI   : CI   :
             :  15MM  :  25MM  : 20MM : 20MM : 20MM : 20MM : 15MM : 20MM : 20MM : 20MM :

 2/4  9 22  stc:177(17) 2 :          :                 : 25  !      :      :      :        :      : 60
 2/4 10 22  stc:197(15) 1 :          :                 :           :      :      :        : 23  1 : 60
 2/4 11 22  stc:           :          :                 :           : 22  1 :        : 23  1 : 60

             :        :        : ABOVE: BELOW: ABOVE: BELOW:      :      : CI   : CI   :
             :  15MM  :  25MM  : 20MM : 20MM : 20MM : 20MM : 15MM : 20MM : 20MM : 20MM :  \

             STATION

             :        :        : ABOVE: BELOW: ABOVE: BELOW:      :      : CI   : CI   :
             :  15MM  :  25MM  : 20MM : 20MM : 20MM : 20MM : 15MM : 20MM : 20MM : 20MM :

 2/6 14 20  stc:        :          :                 :           : 18  1 :      :        :      : 60
 2/6 15 20  stc:        :          :                 :           : 18  1 :      :        :      : 60 •
 3/0  2 20     :        :          :                 :           : 19  1 :      :        :      : 60
 3/0  3 20     :        :          :                 :           : 16  2 :      :        :      : 60
 3/0  9 20     :        :          :                 :           : 17  1 :      :        :      : 60
```

Appendix IA

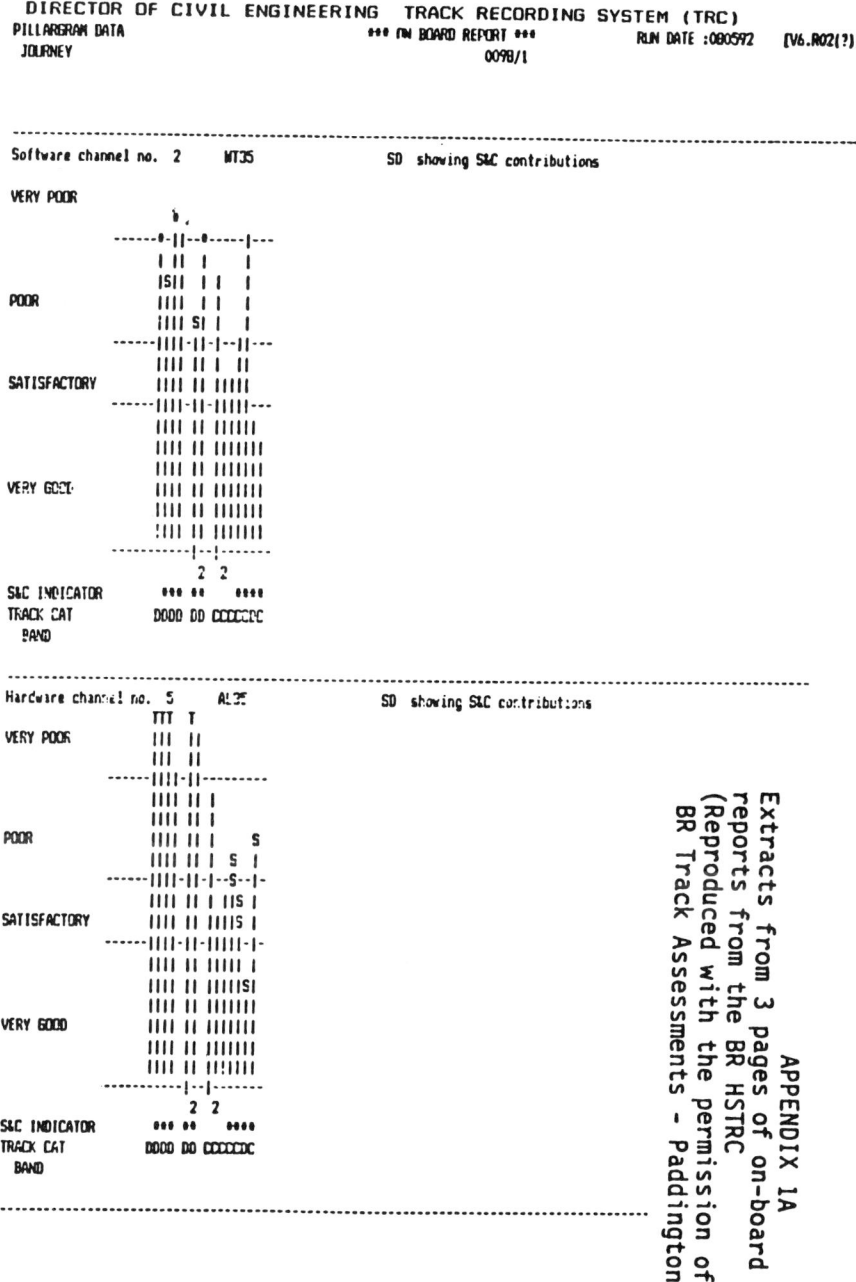

DIRECTOR OF CIVIL ENGINEERING   TRACK RECORDING SYSTEM (TRC)
PILLARGRAM DATA   *** ON BOARD REPORT ***   RUN DATE :080592   [V6.R02(?)]
JOURNEY   0098/1

Software channel no. 2   MT35   SD showing S&C contributions

APPENDIX IA

Extracts from 3 pages of on-board reports from the BR HSTRC (Reproduced with the permission of BR Track Assessments - Paddington)

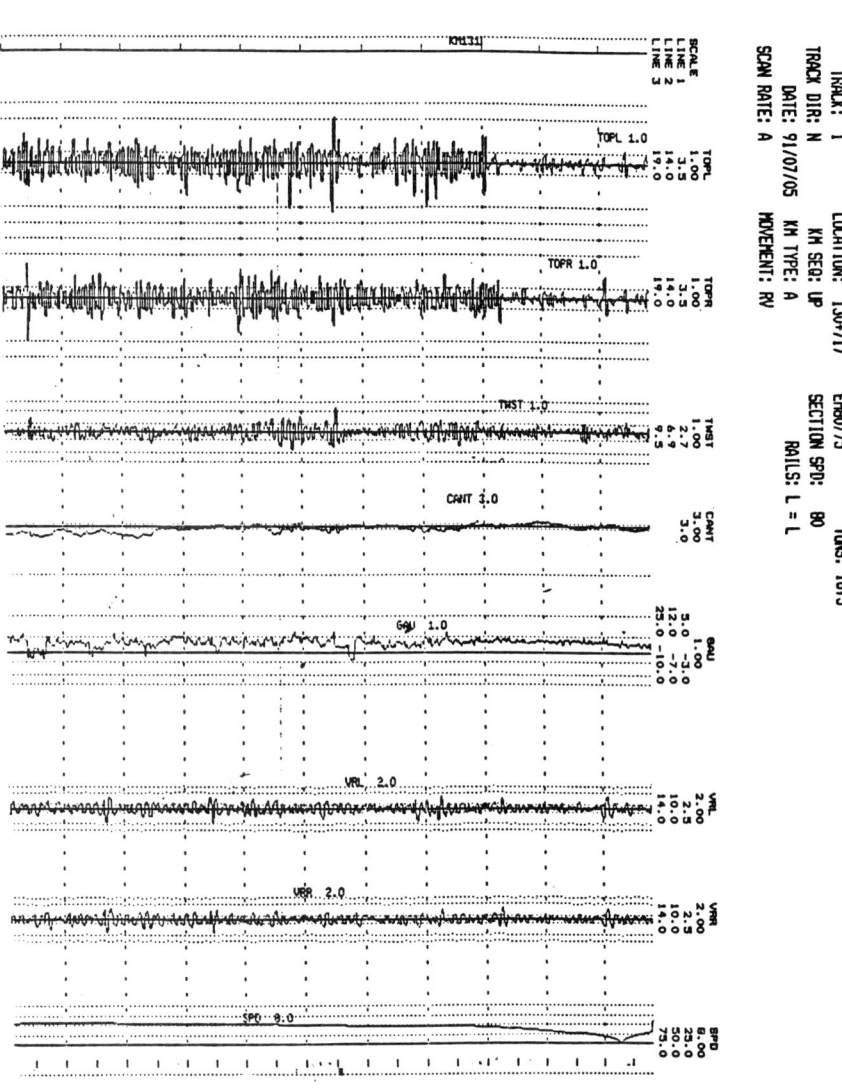

"PLASSERAIL" EM 80

Appendix II

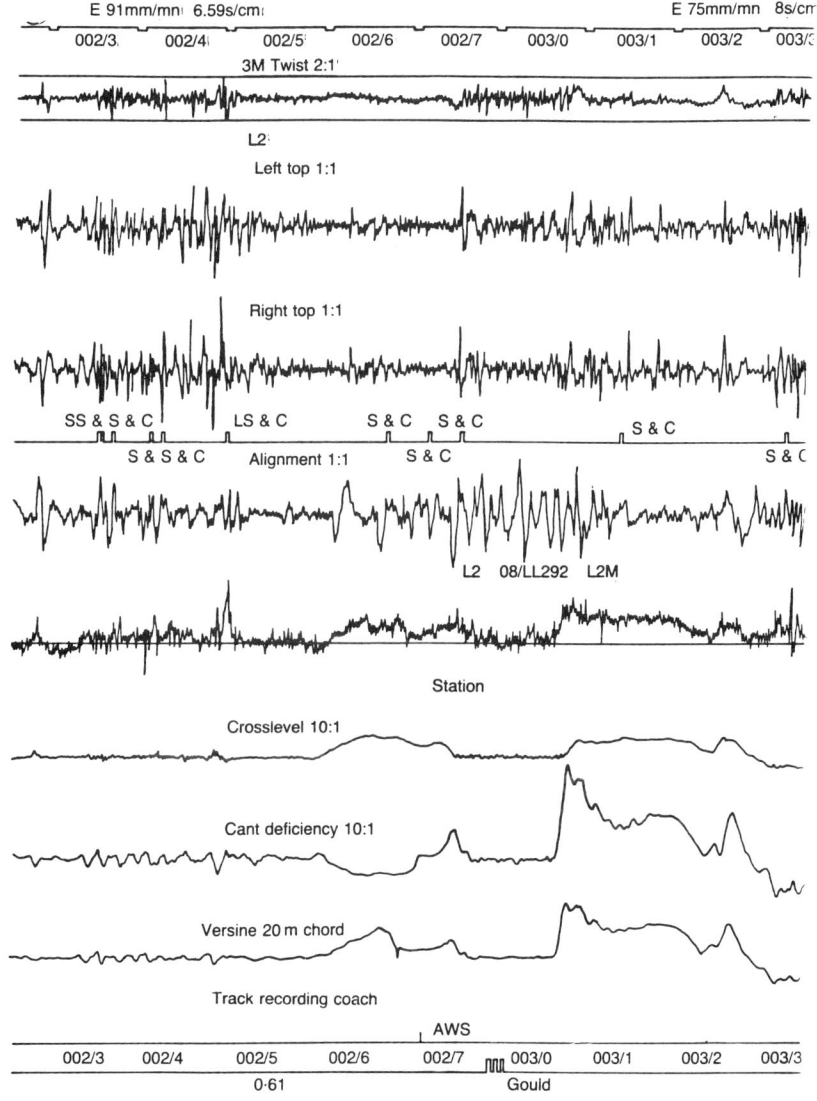

PAGE: 1   DETAIL   REPORT   1991/07/05

STATION: 16,5 TON AXLE LOAD   TO   EM80/75   SECTION NO. M

PARM	------B STANDARD EXCEPTION------			LEN	MAX AT	MAX	------C STANDARD EXCEPTION------		
	KM/M	TO	KM/M	LIMIT	(M)	KM/M	VALUE	KM/H	TO	KM/H	LIMIT
TWST	131+243		131+245	+7MM	2	131+244	+8MM			
TOPL	131+245		131+246	14MM	1	131+245	+15MM			
TOPL	131+248		131+249	14MM	1	131+248	-16MM			
TOPR	131+319		131+319	14MM	1	131+319	-14MM			
TWST	131+756		131+756	14MM	1	131+756	-14MM			
TWST	132+33		132+14	+7MM	2	132+36	-8MM			
TWST	132+36		132+37	+7MM	1	132+39	+8MM			
TWST	132+39		132+40	+7MM	1	132+42	-8MM			
TWST	132+42		132+43	+7MM	1	132+208	+1MM			
TOPR	132+208		132+209	14MM	1	132+987	+18MM			
TOPL	132+987		132+988	14MM	1	133+175	+1MM			
TOPL	133+174		133+175	14MM	1	133+323	-7MM			
TWST	133+322		133+323	+7MM	1					

PAGE: 1   KILOMETRE SUMMARY REPORT (LOT = 50M)   1991/07/05
                         TURNOUT AND SPLICES   ===EXCLUDED===

STATION: 16,5 TON AXLE LOAD   TO   EM80/75   SECTION NO. M

| | 1 | 2 | 3 | 4 | 5 | 6 | 7 | 8 | 9 | 10 | 11 | 12 | 13 | 14 | 15 | 16 | 17 | 18 | 19 | 20 | (0) |

KM 131 TO 131 LOT NUMBER
TURNOUT(//), SPLICE(===)
EMERGENCY WORKING (***)
EVENT   20 (0)

| | 1 | 2 | 3 | 4 | 5 | 6 | 7 | 8 | 9 | 10 | 11 | 12 | 13 | 14 | 15 | 16 | 17 | 18 | 19 | 20 | (1000) |

KM 131 TO 132 LOT NUMBER
TURNOUT(//), SPLICE(===)   TP TP   TP   TP   TP
EMERGENCY WORKING (***)   TW
EVENT   20 (1000)

Appendix 2A

```
PAGE: 1        10 KILOMETRE SUMMARY REPORT   (TURNOUTS INCLUDED)        1991/07/05

        STATION:                TO                          SECTION NO.  M

                        16,5 TON AXLE LOAD        EM80/75
```

AFSTAND DISTANCE		AXLE LOAD	WRINGING TWIST			SPOORWYDTE GAUGE			KANTING CANT			PROFIEL PROFILE			LYNRIGTING LINING			TOTAAL TOTAL			TOTAAL TOTAL
			6.9	I	9.5	12 -7	I	-25 -10	990	I	999	14	I	19	I10.0 I0.01 I10.0	I	+14.0 *0.01 M14.0		I		
VAN I TOT		TON	M	I	N	M	I	N	M	I	N	M	I	N	M	I	N	W	I	N	W+N
130.991 : 140.000		16,5	6	:		0	:		0	:		18	:		4	:		24	:	1	25
140.000 : 141.000		16,5	0	:		0	:		0	:		0	:		0	:		0	:	0	0
TOTAAL/TOTAL			6	:		0	:		0	:		18	:		4	:		24	:	1	25

	M	I	S
	2.78	I	1.30
	0.00	I	0.00

```
GEMIDDELDE AANTAL WERKPLEKKE PER KM VIR DIE TRAJEK (W+N)
AVERAGE NUMBER OF WORKING PLACES PER KM FOR SECTION (W+N)       M  =   2.50

STANDAARDAFWYKING VIR TRAJEK
STANDARD DEVIATION FOR SECTION                                  S  =   2.51
```

Appendix IIA

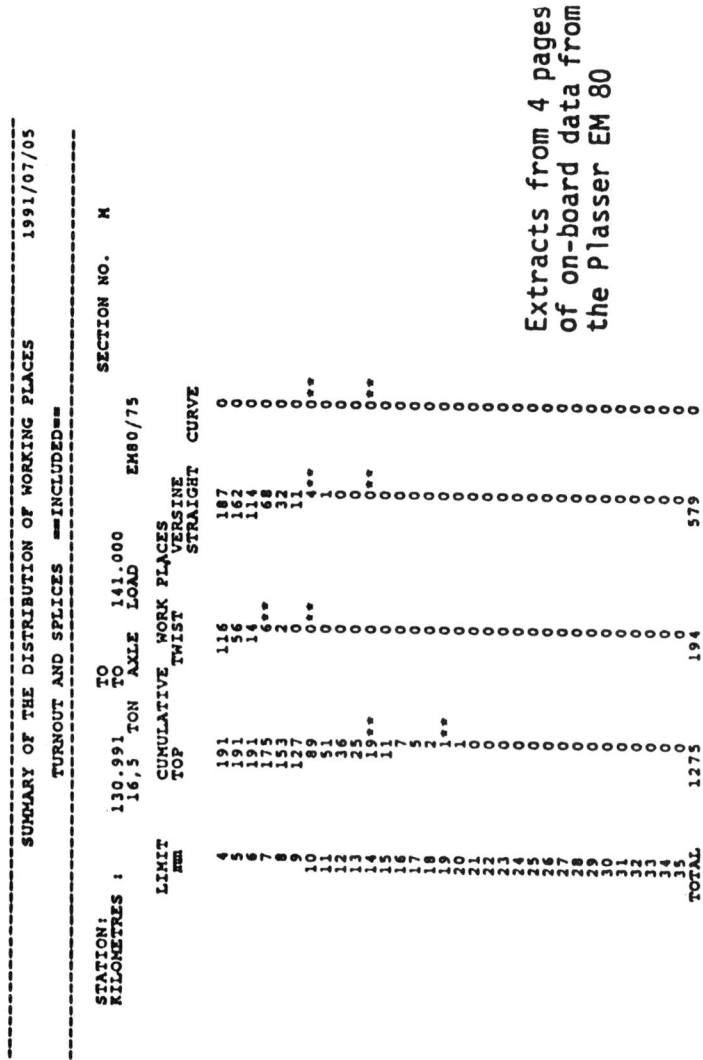

SUMMARY OF THE DISTRIBUTION OF WORKING PLACES   1991/07/05

TURNOUT AND SPLICES   ===INCLUDED===

STATION: SECTION NO.   M
KILOMETRES :   130.991 TON TO 141.000   EM80/75
               16,5 TON AXLE LOAD

                    CUMULATIVE WORK PLACES
LIMIT   VERSINE
mm   TOP   TWIST   STRAIGHT CURVE

4   191   116   187   0
5   191   56   162   0
6   191   14   114   0
7   175   6**   68   0
8   153   2   32   0
9   127   0**   11   0
10   89   0   4**   0
11   51   0   1   0**
12   36   0   0   0
13   25**   0   0**   0**
14   19**   0   0   0
15   11   0   0   0
16   7   0   0   0
17   5   0   0   0
18   2**   0   0   0
19   1**   0   0   0
20   1   0   0   0
21   0   0   0   0
22   0   0   0   0
23   0   0   0   0
24   0   0   0   0
25   0   0   0   0
26   0   0   0   0
27   0   0   0   0
28   0   0   0   0
29   0   0   0   0
30   0   0   0   0
31   0   0   0   0
32   0   0   0   0
33   0   0   0   0
34   0   0   0   0
35   0   0   0   0
TOTAL   1275   194   579   0

Extracts from 4 pages
of on-board data from
the Plasser EM 80

Appendix III

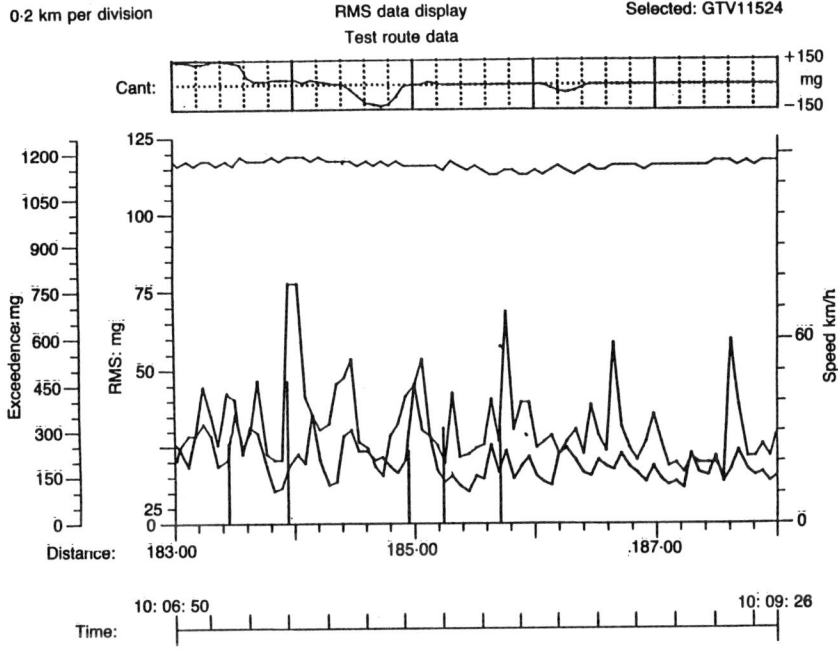

Discussion on Papers 3– 5

H. M. AHMED, former General Manager, Sudan Railways
How can concrete sleepers which are very heavy be layed manually?

Comparing contract work and direct labour, I would prefer contract work by local contractors for small projects and maintenance of track. For large projects I would prefer international contractors because of the transfer of new technology, training of local personnel and procurement of machining and employment which can be used later for maintenance of the track by railway personnel.

F. I. MAU, Vice-President Operations, BHP Rail Products (Canada) Ltd
Steel sleepers have had problems with vibration stability but this problem has been designed out of steel sleepers. 90 km/h operation is normal in Australia and over a steel sleeper test on Canadian national railroad near Toronto, no vibration was evident. Steel sleepers are normally installed and maintained with standard timber sleeper equipment. The principal change in method required between timber and steel sleepers is that more tamping is required with steel sleepers. However, once filled with ballast, the latest box ended, cant in railseat type of sleeper design then has virtually identical maintenance requirements as timber sleepers except that gauge control is eliminated.

By mixing steel sleepers of appropriate design, typically 1 in 4, with timber, a track of uniform characteristics can be produced, while holding gauge and extending the life of the timber sleepers.

R. SILVER, Senior Engineer, Maunsell PB Ltd, UK
I am interested in Mr Seshagiri Rao's development of Top Track and its facility for iterative assessment of alignments.

Whereas there is much discussion that centres around desirable track standards, and these after all have a crucial bearing on any acceptable design realignment for maintenance improvement, I will confine my present comments to those areas of realignment determination that have a basis in computer application.

Mr Seshagiri Rao's Paper raises an important question in respect of constrained alignments in the proximity of fixed infrastructure and the instances of tunnels and girder bridges have been given as examples. It is in

DISCUSSION

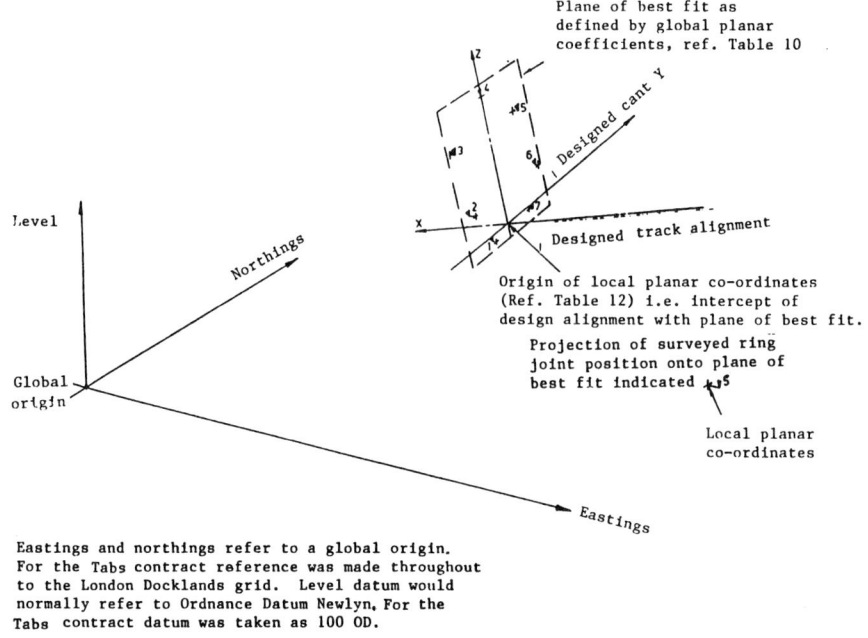

Plane of best fit as
defined by global planar
coefficients, ref. Table 10

Designed cant

Designed track alignment

Origin of local planar co-ordinates
(Ref. Table 12) i.e. intercept of
design alignment with plane of best fit.

Projection of surveyed ring
joint position onto plane of
best fit indicated

Local planar
co-ordinates

Level

Northings

Global
origin

Eastings

Eastings and northings refer to a global origin.
For the Tabs contract reference was made throughout
to the London Docklands grid. Level datum would
normally refer to Ordnance Datum Newlyn. For the
Tabs contract datum was taken as 100 OD.

Fig. 1. Relationship of local to global axes

these areas that track maintenance, without infringement of minimum track alignment criteria is the more difficult in that the idealized lifts and slews of the Top Track alignments may have to be compromised. It is these areas that are of particular interest to me.

A significant proportion of railway in the UK is tunnelled. Historically, the business of railway operation, being a conglomeration of different companies has brought with it an inheritance of amalgamated standards. At the turn of this century the Board of Trade recognized 146 load gauges for the operations of all companies and these varied between 7ft 6 in and 10ft 6 in in width and 9ft and 15ft in height. Of the structures that accommodated these various gauges under their original private ownership many still exist today. This is not to infer that present structure gauges are in any way an inferior compromise, but it does reveal that modern rail vehicles and the predictability of their dynamic performance are pushing structure gauging and the optimization of maintianed track alignments to the limits of measurable accuracy. British Rail, for example, has put in place the infrastructure amendments occasioned by the introduction of trans-Channel Tunnel stock onto Network SouthEast. London Underground, similarly, with its refurbishment of lines is taking proper account of the exisitng tunnel structures in its replacement and maintenance of track and renewal of rolling stock.

Maunsell's introduction of computer assistance to the resolution of alignment of track in tunnels began around 1987 when they, with Edmund Nuttall Ltd, were awarded the design/construct contract for the tunnelled extension of the Docklands Light Railway to Bank Station. At the outset of that contract we recognized the potential for linking the computerized track and hence tunnel, design alignments with the facility to both steer the shield that drove the tunnels and subsequently to identify the wriggled track alignment that could be accommodated within the tunnels as built.

Computer means are ideally suited to rationalizing the mass of data that accumulate from such an exercise, some 8-10 Mega Bytes/km in our experience. The numerical model so assembled is a valuable tool in the assessment of that primary railway asset, the tunnel.

Very briefly, to explain the assembly of the numerical model, conventinal field survey of tunnel sections by total station instrument is data logged directly to the computer program. This, together with the track alignment, either existing or in the case of a new tunnel, as designed, provides the necessary input. Each tunnel section is identified by a plane of best fit which by intercept with the track alignment reduces the 3-D global referencing of the system to a 2-D analysis. With an accurate knowledge of the track alignment as it approaches and leaves the section it is possible to model the attitude and throws of the rail vehicle as it passes through the section. It is a simple matter then to compute the structure clearances. For the DLR the minimum structure clearances to the moving vehicle were 75 mm. On the LUL Central Line Refurbishment Project, where we have completed 50 km of tunnel and track assessment, structure clearances are just 25 mm.

Clearances may be output in tabular or horizontal plot form and it is the latter that gives the means to identify the appropriate wriggled track adjustments. As Mr Seshagiri Rao has correctly emphasized it is very important to consider the entire alignment, rather than just the part that is the cause of the strucutre infringment in order that the alignment is best optimized. A design aligned track will return better operating speeds, higher levels of comfort and safety and lower maintenance costs in track and vehicles.

Our work in this area of computerized track and tunnel assessment has reached the stage where speed profile driven dynamic performance of vehicles is a reality and I would welcome the opportunity to take up with the author the ultimate challenge of fully automated computerized track realignment for the optimization of tunnel space.

Finally I whole heartedly endorse Mr Seshagiri Rao's point that monumented track referenced to an optimized design alignment is the most assured means by which track can be properly maintained.

DISCUSSION

P. EVANS, Principal Permanent Way Design Engineer, British Rail Network SouthEast

British Rail, Network SouthEast are undertaking a clearance survey and adjustment work to achieve clearances for Channel Tunnel Networker and Freight container rolling stock in areas of very tight structural clearance.

British Rail use on-track mechanical and manual surveys and computer programmes to achieve optimum track alignments.

E. CHIMIDZA, Deputy Chief Civil Engineer, Botswana Railways

Is it necessary to deploy patrolmen two to three times a week, on continuously welded track on concrete sleepers?

R. D. SAWYERS, Director, Travers Morgan Railways, UK

Comment was made in Paper 4 and in subsequent discussion on the use of computerized techniques for assessment of clearances in tunnels and refinement of alignments. Although we are track engineers we should not lose sight of the fact that the vehicle kinematic envelope is an equally important factor.

Travers Morgan Railways recognized this fact in 1990 and have developed a computerized technique to derive the true kinematic envelope for individual rolling stock using basic vehicle parameters. In parallel a programme has also been developed using MOSS taking structure gauging data and track alignment information and converting this into a 3D model of the tunnel or other structures.

C. W. COATES, Technical Audit Engineer, London Underground Ltd, UK

Mr Forbes mentioned his responsibilities as setting standards and auditory compliance. Could he explain his means of auditory.

D. R. BALLINGER, Author

In reply to Mr Chimidza, on continuously welded track on concrete sleepers, which is in generally satisfactory condition, it is not necessary to deploy patrolmen two or three times a week, particularly on low speed, low tonnage lines.

With modern track, the need to attend to fastenings and joints, which was one of the main reasons for introducing patrolmen, has been effectively eliminated.

In consequence it is probably sufficient to patrol once a week as an interim inspection between those of the gang leader or permanent way Inspector.

With little or no maintenance work to occupy his attention, the Patrolman will need retraining in the types of things he should be looking for in CWR, such as geometrical discrepancies, component defects, ballast erosion (particularly at bridge corners and illegal crossing points), etc.

Special patrolling during hot weather is, of course, outside any normal inspection arrangements and is usually only necessary where the track is known to be weak (i.e. at risk to buckling) due to recent disturbance by tamping, where ballast profiles are substandard, or where for some reason the track has not been properly destressed. In these situations it will be necessary to patrol throughout the heat of the day, including imposing speed restrictions as neccesary until such time as the condition causing the need to patrol extraordinarily has been rectified.

M. SESHAGIRI RAO, Author

The design of the vertical rail top profile as carried out by the Top Track programme is an iterative process. Human intervention is possible and is even necessary to take into account the large number of constraints and restrictions which are inevitable in an old track. Electrifications, tunnels and overbridges all impose their own limitations.

On the other hand, curve realignment by Top Track is not an iterative method. It is a straightforward solution of diffential equations. The speciality of Top Track is the sophistication of its mathematics. It has no relevence to any special measuring equipment or slewing machinery. The best or whatever is available may be used.

An ideal reailway curve is symbolized by its perfectly trapezoidal versine diagram. For any situation, millions of such solutions are feasible. Most of them would, however, ask for substantial slews incompatible with the existing structural constraints. Top Track software guarantees the solution physically closest to the existing curve, hence needing the least slewing effort. The slews alternate between inward and outward every few stations and their alegraic sum is zero.

It is therefore rare that slewing restrictions come in the way of implementation. We are serving railways of seven countries and have usually been able to give perfect trapezia within physical restrictions in about 95% of the problems sent to us. However, in the remaining few cases, one of the several methods of compromise could be attempted.

One method is to have a compound curve having more than two transitions, say three or four or five. Another would be to have a marginal distortion of the versine diagram by superimposing an exponential easing curve of contra slews. These are illustrated in Figs 1 and 2.

We do not, however, have any experience of solving curves from the British Railways. We are keenly interested in comparing our results with those used on the British Rail Network South East referred to by P. Evans.

DISCUSSION

Mr Silver have very clearly brought out the problems in very old tracks particularly through tunnels. In India we also have the same problems. The vehicle width has steadily gone up over the years from less than 10 ft to 12 ft (3.66 m). The mast clearance in the initial stages of AC electrification was only 2.6 m in the recent electrification projects. It is at such tight locations that the advantage of Top Track in fitting a perfectly trapezoidal versine curve is really appreciated. Compromises, if any, are very nominal.

Not all will agree with Mr R. D.Sawyers that the clearance gauge should be changed from time to time in keeping with the changes in the vehicle designs. It should actually be the other way round. Permanent structures are already there and it is expensive to enlarge them. On the other hand vehicles can always be designed to fit the existing structural clearances.

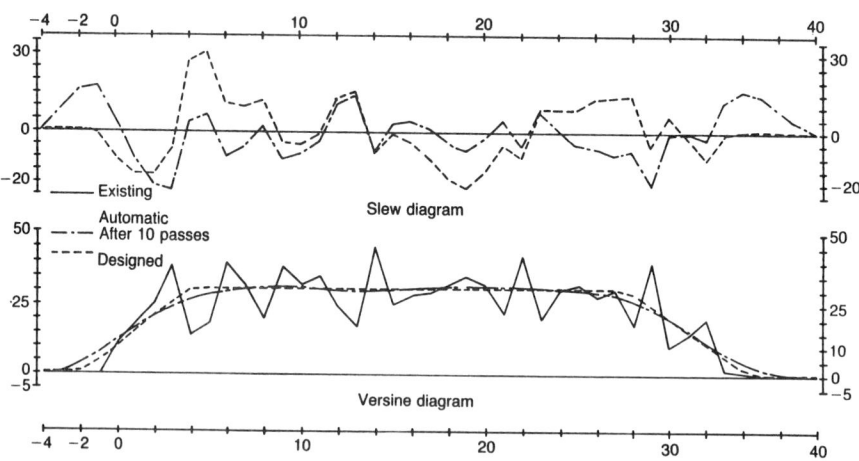

Figures 1 and 2. Curve No. 28

Mr E.Chimidza's query is very relevent to tropical railways. Whereas the temperature difference between the coldest winternight and the hottest summerday is very much more in European countries than near the equator. It does not happen that the propensity to buckling and fracture is more in temperate climates than in tropical. In the tropics the diurnal variation in temperature between day and night is very much more than in temperate climates. Even though the range is smaller, the number of cycles of compression and tension is several hundred times what it is in cold countries. Consequently this creates pockets of high compressive and tensile forces within a long welded rail. it is therefore essential that CWRs in tropical climates should be restricted in length to what can be easily destressed at a time. Destressing operations should be carried out at least twice a year and switch expansion joints provided at reasonable intervals. If these are done, the frequency of patrolling is unimportant.

P. J. FORBES, Author

The normally accepted method of audit uses one/two auditors taking up to three weeks at audit location and a further three weeks to publish the report.

The Regional Railways' civil engineers' method uses ten knowledge-able/experienced members of his management team as auditors. The auditee nominates an equal number of audit co-ordinators whose responsibility it is to provide verification, as required during the audit.

This method takes two days to complete an audit and causes little disruption to the day to day activities of the audit location.

The results are analysed and the report published within one week of audit.

6. Bridges and structures

M. ARSHAD, BSc, FICE, Principal Civil Engineer, Transmark and
J. H. G. Cook, MICE, FCIT, Consultant, Transmark

Introduction

The scope of the present conference is track maintenance, and it would therefore not be appropriate to give detailed consideration here to the other major element of the railway civil engineer's field of responsibilities — bridges and structures; that would be a sufficiently large subject for a separate conference.

However, there is a vital inter-relationship between the track and those structures which support it or cross it, and it is necessary for staff at all levels in the track maintenance department to have a reasonable understanding of them - their purpose, the way they function and their requirements in terms of examination, inspection and maintenance. Conversely, the designers of bridges should have a clear understanding of the requirement of those responsible for track maintenance, in terms of giving support for the track in a manner which allows good track geometry to be achieved and ease of maintenance. The thrust of this Paper is therefore directed at the implications and interactions of bridges and structures with the track, and to the staff who maintain it.

Bridges comprise the major part of the structural engineering of most railways. The majority of railways, particularly those built to carry comparatively slow moving and infrequent trains, have only a few bridges other than those which carry the line over significant water courses. In exceptional cases or in particularly hilly terrain there may also be viaducts built to avoid the construction of large embankments, or to span gorges. In densely populated areas there may also be a few road bridges, usually over the line. It is also the general practice for new major roads to cross any railway, large or small, by bridges.

At the other end of the scale, the railways in the UK have a large number of bridges, averaging about two per kilometre, mostly carrying roads or tracks over the line, or vice versa. In the major urban areas, long stretches are raised on continuous viaducts.

This is a situation which arises from the complexity of the road network in the UK, the density of traffic, both road and rail, and the limited numbers of level crossings that were permitted. The provision of bridges thus varies widely from one railway to another.

Railway structures can be classified under the following headings

- culverts
- underbridges
- streams, rivers, canals, other railways, creeks, wadis
- roads and cattle creeps
- viaducts
- overbridges
- footbridges and passenger subways
- tunnels

Other structures

- station buildings
- other buildings (workshops, depots, signalboxes, staff accommodation, housing, etc)
- platforms and docks
- retaining walls
- river training and protection works
- masts (for lighting, electrification etc.)

On some administrations, the railway engineering department also has responsibility for sea defence works, and sometimes docks and harbours. Nevertheless, the major maintenance liability is nearly always in the bridges.

Design of new, strengthened or reconstructed bridges

Cost-effectiveness in bridges and structures begins with good design, and good design starts with an understanding of the local conditions in which they are to be built and have to exist. The track maintenance staff, with their day-to-day close contact with the track and observation of the local environment of the railway, are the front line of the railway civil engineering workforce. Their knowledge of and presence in the local scene must be made full use of in all aspects of the bridge engineer's task, not just for maintenance but from the initiation of new or reconstructed bridge works on an existing railway as well.

Even at the early stages of the design of new, reconstructed or modified bridge works, the designer should visit the site if at all possible, and the track staff should be consulted. Ground survey work of various kinds topographical, geotechnical etc. will be undertaken by specialists, but the designers would do well to listen to what the local gangers and supervisors have to say, particularly if they may have a knowledge of the area going back some years. Their reports are likely to be only anecdotal, but accounts for example of what happened on previous occasions of flooding and changes in the hydrological regime that have taken place can be a very valuable adjunct to quantitative scientific survey.

In the design of new, reconstructed or altered bridges the designer should pay particular attention to the future requirements of the track maintenance staff in order to allow them good access for inspection and maintenance purposes. This is an area which can be overlooked or given inadequate attention, and often in the past proper provision for walkways, handrails and refuges has been widely neglected. There are very many bridges throughout the world which are unsatisfactory in meeting basic requirements for the track maintenance staff who have to work on them, not only failing to provide proper safety standards, but inhibiting easy access and facility for track maintenance staff to carry out their work. Walkways, handrails, refuges, and in the longer bridges room to locate plant should form part of the design brief.

Underbridges are there to support the track properly, and that means to allow good line and level to be achieved and maintained over it and through its approaches. Normally for short span bridges (say up to 30 m), on lines where trains of medium or high speed are to be operated, the most satisfactory way of achieving this is to carry the same form of track construction over the bridge as applies to either side of it. In the great majority of cases this will be ballasted track. In longer bridges it may be calculated that the extra deadweight of ballast will make this form of construction uneconomic, and some form of direct track fastening may be adopted. From the point of view of the track maintenance staff, however, the ease of maintenance of ballasted track will nearly always lead to a preference for this track form.

Whether the track over the bridge has ballast or not, particular care should be taken in the design to ensure that the transition of the track from the supported to the unsupported condition should be smooth, by, for example, the provision of 'running on' slabs under the track at the ends of the bridge.

There is a tendency, particularly in some developing countries, to reconstruct bridges purely because they were designed to a certain loading, and since they may have deteriorated slightly they are now considered unsuitable to carry the loadings planned for the future. The fact of the matter is though, that in many cases the actual strength is not known.

It should go without saying that before reconstruction is decided upon the condition of the bridge should be properly measured and the structure subjected to a critical technical assessment to ascertain its true carrying capacity. Experience, however, shows that this is not always done although there are well established techniques for making such assessments for all the principal types of bridge design, including masonry and brick arches.

If the bridge is calculated to be weak, then the next step is to consider strengthening, which, although this is very often a much cheaper alternative than reconstruction, is not always treated as a serious choice. All feasible options should be calculated and costed; the price of doing so is small compared to the financial penalty of selecting the wrong one.

It has to be said that in some cases the decision to reconstruct is taken for reasons other than engineering efficiency. It may be perceived that there is greater credit to be gained from designing and building a new bridge than in patching up an old one, although there is often more engineering skill in the latter achievement. Furthermore, in the past at any rate, funding agencies have been more ready to invest in new works than the repair of old ones.

In certain cases local practical reasons may require that bridges be reconstructed if they are found to be inadequate. For example, in Uruguay the many metal bridges cannot be easily strengthened or repaired because of the lack of local capacity to effect alterations to structural steelwork. Bridge designers, sometimes working in a different country, should be careful to appraise themselves and take full account of local factors such as these.

Modern techniques of life-cycle costing should be used in the evaluation of alternative designs, although limited available funds often make 'minimum first cost' the criterion particularly in the developing world. This unfortunately means that considerations of limiting the future maintenance costs have to be sacrificed in the interests of minimising initial capital outlay.

Whatever the financial pressures on him, the responsible bridge designer should be mindful of the future maintenance requirements of the bridges which he has to engineer. In many cases, provisions can be made at the design stage, at quite modest cost, which can save major maintenance difficulties and expense in the future.

Certain bridge components may well have a shorter life than the structure as a whole, and provision should if possible be made for the easy replacement of these components. For example, a concrete bridge with a possible life of 120 years may have bearings made of a composition material with a life of only 20 years. This can be sensible and economic engineering provided that the bridge has provision in its design to be easily jacked up to facilitate the replacement of its bearings from time to time.

The choice of materials of construction has, of course, an important influence on the future maintenance liability. Concrete — reinforced or prestressed — is normally the preferred material, partly because of its durability. However, problems are being experienced in parts of Europe with serious deterioration in concrete bridges arising from de-icing salts.

In wet climates the need for good waterproofing systems and detailing is now well established with bridge designers. But even in comparatively arid areas, there is s requirement to consider waterproofing at the design stage. Wind-blown sand often contains corrosive salts which attack metalwork and sometimes concrete when it is moistened by rain. The waterproofing should be robust and if necessary protected from damage by track maintenance staff, by means of a layer of tiles, for example.

There is a tendency in some countries to let turnkey contracts for construction of bridges. This type of contract does not necessarily produce bridges which require low maintenance. It is also difficult to ensure that the contract properly covers all the required elements. even at the construction phase. As a general principle, it is advisable to prepare a detailed design before inviting tenders, allowing the client to ensure that proper consideration to maintenance aspects is taken into account at the design stage.

Installation of bridges

Information from the local maintenance staff should be sought and can be helpful in deciding access arrangements and the most appropriate time of the year for carrying out major bridge works.

Normally the installation of the track over a new or reconstructed bridge will be undertaken by the contractor or a relaying gang. At the completion of the work there should be a formal handing-over arrangement for the track on the bridge to the maintenance staff before the latter assume responsibility for it. At no time should there be any doubt about who is responsible for the maintenance of the track, although the local permanent way staff should always have the final responsibility for authorising the passage of trains over the line.

As far as possible track joints and points and crossings should be kept off bridges.

Examination and inspection of bridges and culverts

The achievement of cost-effectiveness in the maintenance of railway bridges and engineering structures depends primarily on an efficient system of examinations and inspections, followed by strength assessment, the diagnosis and prognosis of problems, the development of appropriate measures to deal with such problems and the correct prioritisation of maintenance works. The cost of achieving all this is small compared with the benefits it can bring. It has to be remembered that the only purposes for planning to spend money on any bridge are

(a) safety (of traffic, staff and third parties)
(b) meeting the traffic requirements
(c) prevention of heavier expenditure later on
(d) aesthetics.

On well-managed railway administrations there will be a complete procedure for the systematic examination and inspection of bridges. A typical railway will have a chief civil engineer with overall responsibility for the safety and efficiency of the civil engineering infrastructure. Under him there will be a designated engineer with the task of setting bridge engineering standards and undertaking design, and district engineers with the responsibility for maintenance and inspection of bridges. The district engineer will have on his staff a specialist bridge engineer, works supervisors and bridge examiners.

Engineer's inspections on such a railway will normally be related to the scheduling of a programme for bridge repairs, and the preparation of a budget for works to be performed in a year or two's time. To ensure that expenditure is cost-effective — that is to say that work is done efficiently, in the right order of urgency and not carried out earlier than is necessary — a system of priorities will be employed. In any case, before any except emergency work is put in hand, the justification for it will need to be agreed with the business managers (to use the modern jargon) of the line.

British Railways have developed a computerised bridge management system called BASIS (Bridges and Structures Information System). This records various data about the structures, including

(*a*) geographical location
(*b*) type of structure
(*c*) number and size of spans
(*d*) material of construction
(*e*) date of construction and any modification
(*f*) its carrying capacity - permitted axle load and speed for railway bridges
(*g*) weight limit for road bridges
(*h*) name of local authority
(*i*) engineer responsible for inspection and maintenance
(*j*) date last inspected and date of next inspection.

This system, among other things, produces a list of structures which are due for inspection in a certain period, and will also produce a list of maintenance work which is overdue. Similar systems can readily be set up without great expense on even quite small railways.

In the United Kingdom, the Department of Transport has devised a simple but effective way of recording the works required to be undertaken to highway bridges, which helps in the planning, prioritisation and budgeting of bridge maintenance works. The basic data for each bridge is recorded on a single sheet, as shown in the example in Appendix I.

On British Railways, the frequency at which various examinations should be undertaken is set out in the *Civil Engineering Handbook No 6 - Examination of Structures*. An extract from this is given in Table 1.

The periodicities for the various examination tasks are a matter of engineering judgment, and need to be reviewed from time to time. Different administrations will reach their own conclusions as to what are appropriate for their particular cases. On BR the frequency of bridge examiners inspections has changed from every three years, then to four, and currently it is six-yearly.

Table 1. Frequency of examination

Structure	Type of Examination	Maximum Interval Between Examinations	Examination by
Bridges and Structures Bridges, culverts, retaining walls, etc. and signal structures	Routine observation Quarterly observation	During normal track inspection 3 months	(Patrolman (PW Supervisor PW Supervisor
	Superficial	1 year	Senior Supervisor or Section Works Supervisor or Appointed Examiner
	Detailed	6 years	Appointed Examiner
Bridges & structures for which another Statutory Authority is responsible	Routine observation Superficial Detailed	During normal track inspection 1 year	)) As above) Responsible Authority
Structures on Closed Lines Sensitive structures	Superficial	1 year	Supervisor or Appointed Examiner
All structures	Detailed	6 years	Appointed Examiner
Special Structures	Superficial	1 year	Senior Supervisor or Section Works Supervisor or Appointed Examiner
	Detailed	To be agreed with Regional Civil Engineer	ACE or his representative with Appointed Examiner
Tunnels	Routine observation Detailed	During normal track inspection 1 year	(Patrolman (PW Supervisor ACE or his representative with Appointed Examiner
Tunnel Shafts (including projections above ground level)	Superficial Detailed	1 year 6 years	) ACE or his representative) with Appointed) Examiner

It will be seen that in respect of bridges and tunnels the track maintenance staff have certain responsibilities; in fact, in most cases the bridge and structures are only visited by works staff at yearly intervals. Between these annual visits the track staff generally have the sole responsibility for examinations of bridges and tunnels. Whilst these examinations can only be superficial, they are clearly very important in the routine monitoring of the behaviour of structures.

The procedures and responsibilities vary from railway administration to another. In some organisations, particularly in the developing world, they may be less well-defined and structured. Nevertheless, almost universally, whilst principal responsibility for the examination, inspection and maintenance of bridges rests with specialist staff in the works department or its equivalent, the local track maintenance staff have their vital part to play. Essentially that role is to be watchful for any changes which may take place to the bridge, day by day and week by week, or to factors which could affect it, and take any immediate action that is necessary.

In most cases the action to be taken by local track staff in respect of bridges and culverts will be confined to reporting through the recognised channels or summoning specialist bridge supervisors, should this appear to be necessary. In some cases where faults come to the attention of the track staff it will be necessary to take immediate action without reference to others, such as the imposition of a speed restriction or even closing a bridge or the railway, if the local responsible member of the track department in his judgment sees it to be necessary.

It is not uncommon, particularly in arid parts of the world, for major damage to occur to bridges as a result of sudden storms and the resultant very fast run-off. In North Africa and the Arabian peninsular, for example, this phenomenon is probably the most frequent cause of serious damage to bridges. In Great Britain possibly the most frequent serious cause is damage by road vehicles. In both cases, it will normally be the local track maintenance staff who will be the first responsible railway personnel to be alerted.

Of course, it has to be appreciated that patrolmen, gangers, track chargemen and track supervisors cannot be expected to have a detailed understanding of structural engineering, and that the instructions given to them in respect of the examination of structures must be clear and simple. On BR the responsibilities of the track maintenance staff are set out in *Civil Engineering Handbook No 6*. These are specific to BR and not totally applicable to other railway systems, but they form a good basis for formulating instructions for other railways, specific to each system.

Factors to be taken into account in respect of bridges and culverts can be wide ranging, and the ganger, patrolman or track chargeman should regularly check the following

(a) the general track line and level over the bridge and its approaches. It is often change in the track geometry over s bridge or its approaches which is the first indication that there is some form of deterioration in the structure

(b) behaviour of the bridge under traffic, including specifically

(i) the liveliness of the structure and of individual girders
(ii) any movement at the bearings
(iii) any swaying of the structure or excessive deflection
(iv) any relative movement between components of the structure, for example between longitudinal timbers and the rail bearers or trough girders that support them.

One of the most important tasks a Patrolman or Supervisor can carry out is the observation of bridges under traffic, which is when structural defects are most likely to be noticeable. Particularly on lines with only a few trains a week, it may be that the bridge examiner and works supervisor when they make the irregular visits, do not have the opportunity to see a train cross the bridge, especially if their own inspection vehicle is stood on the track during the inspection, blocking the passage of other trains!

Track supervisors should as a matter of course ride in the locomotive cab and discuss with train drivers the condition of the track generally and their experience of the behaviour of trains when passing over bridges.

Changes in the traffic pattern on the line can have important consequences on bridges. The track maintenance staff may not be advised of such changes being planned or implemented; the first they know of them could be that there are suddenly more trains running, or they are longer or heavier. On well managed administrations this revised traffic pattern will have been carefully considered and authorised by the civil engineer in the light of the capacity of the track and bridges to accommodate it. However, the affect of such increased traffic on the bridges should be observed, as far as is practicable, by the local track staff, and reported on.

Proper clearances at bridges should be maintained. It seems obvious that these can be detrimentally reduced laterally by realignment and vertically by lifting the track, but cases have occurred of impetuous improvements being made to line and level without due regard to the maintenance of clearances, even on apparently well-managed railways .

Track staff should also be instructed to report any activity by outside parties on land adjoining the line which might affect railway structures, however indirectly. Cases have occurred where farmers have seen the railway as a free dam to retain flood water for irrigation purposes and even constructed unauthorised sluices at culverts to control the flow of water. In another instance, flood openings through an embankment were partially walled up to provide cattle pens. Interference with the natural hydrology can have disastrous consequences at times of flood, and it is the track staff's responsibility to see that such events do not occur on or near the railway, and if they do to report them.

Track maintenance staff should regularly carry out a visual inspection of the underside of bridges and culverts to observe any changes in the condition of the structure and of the state of the watercourse and the land round about. They should familiarise themselves with all the elements of the bridge and look for changes. Cracks or spalling in abutments or wingwalls, leaning abutments, bulging spandrel walls, erosion of joints, broken bearings or sagging girders are examples of the sort of faults which can occur suddenly and should be easily noted by anyone with diligence and common sense. If there is a sudden worsening of the track alignment or level over the bridge or its approaches the ganger or supervisor should not fail to look under the bridge to see if the cause is there .

With structures over watercourses it is particularly important to examine them after heavy flows have been experienced. The foundations of bridges can become underscoured, culverts become blocked and rivers and streams change course, with potentially harmful results to the railway.

In some bridge locations, training walls extend for some way up-stream from the bridge, and in certain instances on the downstream side as well. These works are normally in the ownership of the railway, but whether they are or not the same level of inspection should be given to them as to the bridge or culvert itself.

Often the larger culverts and bridge openings over wadis are used by road vehicles, particularly those which are dry in the normal course of events. Cases have occurred of farmers and building contractors lowering the ground level under such bridges to allow higher vehicles to pass through. This is potentially dangerous, as it effectively increases the abutment height and even exposes the foundations, thereby weakening the structure.

A growing problem at road underbridges is the phenomenon of 'bridge-bashing' when bridges are struck by road vehicles or the loads they are carrying. This can result in the bridge being damaged or even the superstructure knocked off its bearings. The track maintenance staff should be on the lookout for this problem and be on the look out for it in making their inspections.

The most important elements of the responsibilities of track maintenance staff in respect of the examination and inspection of bridges and culverts can be summarised in the following instructions to track supervisors

- be regular and systematic
- use common sense
- report clearly any changes that occur

- do not assume that because a matter has been reported once it can be forgotten about. Repeat the report, and again if necessary, until it is dealt with or a satisfactory explanation is given.

A system of making ex-gratia payments to track staff for the prompt reporting of bridge defects could be appropriate in certain circumstances.

Maintenance of bridges and culverts

The maintenance responsibilities of the track department in respect of bridges is largely confined to the following areas

(a) the track over the bridge and the interface between the track and the bridge structure

(b) The clearing away of rubbish such as sand drift, vegetable matter and debris

(c) in some cases keeping in good order the walkways over the bridge

(d) the cleaning out of culverts.

Where there is ballasted track over the bridge it can be maintained in more or less the same way as the track on either side of it, although the use of on-track maintenance plant tampers and liners may well be restricted because of the design of the structure. Care should be taken in working on the track on bridges not to damage the bridge structure, especially its paintwork and waterproofing system.

Where there is ballasted track over a bridge it may be considered desirable that the sleepers should be of timber in order to limit the deadload and to provide extra resilience between the track and its support. In any case timber sleepers, rather than concrete, should be used on bridges where the depth of ballast under the sleepers has to be less than 200 mm, in order to provide this extra resilience between the track and the structure and to limit the grinding of the ballast between the underside of the sleepers and the bridge deck.

More often than not, however, particularly on lightly used lines or those with comparatively slow-moving trains, the method of support given to the rails over bridges is quite different from that elsewhere on the system, usually entailing rails being fastened to lateral or longitudinal timbers which are directly supported by the bridge structure. Such direct fastening systems present problems to the track maintenance staff in the upkeep of the rail fastenings and the achievement of good standards of alignment and level, particularly where train speeds apply which require high quality track geometry.

The most difficult area for maintaining correct line and level can be at the ends of the bridge, where the support for the track goes from 'hard' on the bridge to 'soft' off it, and where there will be longitudinal stresses in the track induced by the thermal expansion of the bridge structure. Where chronic difficulties in this regard are experienced it may be advisable to provide some structural alterations to mitigate the transition arrangement. Whether expansion switches should be provided at these locations, where the track has continuous welded rail, is a matter upon which there are differing views among railway engineers.

Where there are direct fastening systems for the rails the supervisors should ensure that the local maintenance staff are familiar with the correct maintenance procedures for them and that they are equipped with the right tools for carrying them out.

The regular cleaning out of culverts is an important function of the track maintenance staff, in order to ensure that they are capable of taking flow up to their full capacity at times of storm. After rain or sand storms culverts can become seriously silted up or blocked by floating debris or sand drift, the clearing sway of which can be quite a major operation. If necessary extra staff and heavy equipment should be brought in for this purpose without delay.

Keeping the bridge clean has an importance far greater than is often realised. Accumulations of ballast, sand, soil and vegetable matter can quite rapidly lead to corrosion particularly in such areas as bridge bearings and in odd corners behind gusset plates. BR regulations require that the permanent way supervisor arranges for such cleaning to be carried out once a month. It is doubtful if such a frequency is often achieved, but on any railway it is suggested that thorough cleaning out of the superstructure of bridges should be done at least every three months.

The instructions given to track maintenance staff in respect of bridges and structures should make it clear what they should not do as well as positive actions they should take. Over-enthusiastic track supervisors in remote locations may decide that a bridge needs to be propped, and have been known to take such action without reference to the qualified bridge staff. This is likely to be at best ineffective and at worse disastrous. Bridges built of prestressed concrete, for example, can be catastrophically damaged if wrongly propped.

Initiatives by the track department can sometimes play an important role in the restoration of the line, albeit temporarily, for example when a bridge is washed away in a storm. In the Sudan the track staff are adept and surprisingly quick in building a temporary diversion of the line across a wadi after such an occurrence.

Finally, it should not really be necessary to state that proper clearances at bridges should be maintained. It seems obvious that these can be detrimentally reduced laterally by realignment and vertically by lifting the track. However, experience tells that elementary considerations of this sort are sometimes completely overlooked. Both lateral and vertical changes in track geometry need to be particularly carefully considered on lines which have overhead electrification, and the electrical department consulted.

Other structures

The inspection procedures for tunnels, as far as the track maintenance staff are concerned, should largely follow those for bridges. The staff should be equipped with powerful enough lamps and torches to be able to see clearly the condition of the tunnel structure. On most administrations, tunnels are subjected to regular and rigorous technical inspections from a special train.

In making their inspections of the line, the track maintenance staff should familiarise themselves with the condition of other structures which adjoin the track and note and report any changes which may take place, particularly those that affect the stability of the line or clearances.

Retaining walls, particularly those at the foot of embankments, often tend to get neglected by both works and track maintenance staff. Frequently they become overgrown and it becomes difficult for their condition to be assessed. It is normally the responsibility of the track staff to keep such structures clear of vegetation, not only in order to allow them to be examined, but also to prevent roots penetrating and damaging them.

Platform walls, where they exist of substantial height above the track, should be regularly gauged by the track maintenance staff. It can be helpful if a simple platform gauge is kept for this purpose at each station where there are such platforms.

Training

Obviously it is necessary for works supervisors and bridge examiners to have appropriate training to cover the full range of their responsibilities. Most administrations will have established courses covering the subject for these categories of personnel. What then should be done for the track maintenance staff to instruct them in their limited but important responsibilities in respect of bridges and structures? In the Authors' experience there does not appear to be generally a great deal of consideration given to this.

With regard to patrolmen, track chargemen and gangers, most of whom are likely to have had limited general education and have learned their jobs by experience hands on, classroom tuition is unlikely to be of much benefit and difficult to organise, while written instructions may be hard for them to understand or interpret. A better option could be for them to be required to accompany the bridge examiners on their inspections and for the examiners to give them some on-the-job guidance in their responsibilities in this field. The bridge examiners would themselves need to be trained to do this.

Track maintenance supervisors on many administrations do have residential courses arranged for them, and some instruction on their duties in respect of structures may be given at these. Again, though, some on-the-job instruction in respect of particular bridges in their sections could be given by the works department's examiners and supervisors.

Track maintenance engineers will in the normal course of their training have received some formal education in respect of structural engineering and generally this should be sufficient to provide a basic understanding of the subject. However, they should from time to time be given written information and instruction to ensure that they are kept informed on developments in the subject.

References

1. Civil Engineering Department, British Railways Board. *Examination of Structures.Civil Engineering Handbook No 6*. Revised August 1984.
2. Civil Engineering Department, British Railways Board. *Assessment of the Live Load Carrying Capacity of Metal Underbridges (BR36840)*. Revised June 1983.
3. Department of Transport. *Assessment of Highway Bridges and Structures* Advice Note, HMSO, 1984.
4. Department of Transport. *Highway & Traffic Departmental Standard BD 34/88:Technical Requirements for the Assessment and Strengthening. Programme for Highway Structures. Stage 1: Older Short Span Bridges and RetainingStructures*. HMSO, 1988.
5. The following UIC leaflets are relevant:
   717-1 *Laying of track on ballast on a steel deck* (2nd edition 01-01-80)
   717-2 *Laying of track on a reinforced concrete deck* (2nd edition 01-01-80)
   717-3 *Steel bridges - Laying of track on steel without ballast - Direct laying* (lst edition 01-01-67)
   776-3 *Deformation of bridges* (lst edition 01-01-89)
   777 *Measures for the protection of railway bridges against impact from road vehicles and for limiting the damage caused* (lst edition 01-07-79)
   778-4 *Defects in railway bridges and procedures for maintenance and strengthening* (lst edition 01-07-89)
   779-1 *Recommendations for determining the carrying capacity of existing metal structures* (lst edition 01-07-86)

Appendix I. Structure Assessment Report

Structure Key [5][4][3][2][1]

Structure No [1][9][A] / [M][5][6] / [W][S] / [1][3][4]·[5][0] / [Q] / [1]

Structure Name [L][O][W] [E][B][B]

Region Code [9][9][0][0] Region NSRO

Agent Code [4][9][0][0] Agent Name BARSETSHIRE

Consultant Code [7][9][0][0] Consultant Name BLOGGINS AND SNOOT

PROPOSED REMEDIAL MEASURES

1. Strengthen Superstructure [Y]
2. Reconstruct Superstructure [N]
3. Strengthen Substructure [Y]
4. Reconstruct Substructure [N]
5. Strengthen Independent Retaining Wall [N]
6. Reconstruct Independent Retaining Wall [N]

AFFECTED SPANS

7. Number Of Spans To Be Strengthened [0][5]
8. Number Of Spans To Be Reconstructed [0][0]

INTERIM MEASURES

9. Required [Y]
10. Impose Weight Restriction(tonnes) [2][5][0][0]
11. Impose Lane Width Restriction [N]
12. Impose One Way Working [N]
13. Prop Structure [N]
14. Close Structure [N]
15. Provide Temporary Alternative [N]
16. Monitor Structure [N]

ESTIMATED RESOURCES FOR PROPOSED REMEDIAL MEASURU

17. Estimated Cost Of Remedial Measures £ [ ][ ][5][0][0][0][0]

SIGNATURE OF TEAM LEADER *B. Smith*   DATE 7-MAR-1990

SIGNATURE OF CHIEF OFFICER OR PARTNER *H. Jones*   DATE 20-MAR-1990 SIGNATURE OF DIRECTOR(DTp) *F. Baker*   DATE 7-APR-1990

103

7. A review of the effects of natural damage

D. G. SPERRING, BSc, MICE, Associate, Railways,
Mott MacDonald Group

This Paper undertakes a review of some of the natural hazards to a railway, the likely consequences, and what can be done to minimise the impact. Response to the exceptional, once in a lifetime occurrence is considered as well as the preventative action that may be taken to ameliorate the effects of more frequent disruptive meteorological conditions.

Introduction

Exceptional natural hazards are unavoidable, but engineers, with fore-sight and resources, are able to prepare for such occurrences so that if and when they do happen, the response is quick and effective, to minimise the consequences. With mobilisation of comprehensive assessment of the situation the engineer is able to begin planning repair and reinstatement of the track, quickly and effectively.

To minimise the effects of wind and storm, which occur more frequently, provision may be made in design, where it is economical to do so, as well as through efficient response. In the following sections the effects of some natural hazards are reviewed and a policy for response is proposed, together with, where appropriate, preventative action by design.

Exceptional natural hazards – earthquakes, hurricanes and volcanic action

Although some areas of the world are known to be liable to exceptional natural hazards, nevertheless their occurrence on a year-by-year basis is not predictable, and forecast warnings give little opportunity for action that would effectively reduce damage to property and equipment.

Procedures are usually well established for minimising risk to life, such as curtailing or suspending train services and enhancing identification of localised risks by emergency patrolling, and are implemented as soon as warnings are received. The railway system may also offer or expect to be called upon to assist or supplant the emergency services in sustaining life. In so far as damage to property is concerned the railway engineer must respond quickly and efficiently to restore the infrastructure of the railway so that train services may be resumed.

The Author is fortunate not to have personally experienced the effects of an exceptional natural hazard. To illustrate the effects and the response, extensive reference has been made to an account of an earthquake at Edgecumbe, New Zealand, which appeared in the *Journal of the Permanent Way Institution* (ref. 1).

Earthquake

On 2 March 1987 an earthquake measuring a maximum 6.3 on the Richter scale hit the Eastern Bay of Plenty in the North Island of New Zealand. By sheer good fortune no one was killed and only two people were admitted to hospital with relatively minor injuries. Extensive damage was suffered to hospitals, industrial premises and homes. Water, electricity and sewage services were severed in many places. Telephone communications were seriously affected. Radio telephone alternatives were used but coverage was not complete. Nevertheless the first report of damage was received at the District Engineer's Office just 14 minutes after the first tremor struck.

The railway track was misaligned and had subsided in various places. The worst was a misalignment of 2 m with a drop in level of 1 m. Embankments adjacent to the ends of bridges had subsided by as much as 600 mm relative to the bridges whilst the bridges remained intact. In one section of railway some 12.5 km was affected by numerous misalignments, subsidence and uplift. Three freight trains had been in section. One could be recovered by rail but two were trapped. One train, loaded with perishables was offloaded and the goods delivered by road. Despite the extensive damage, the track was returned to service by Wednesday 11 March, just nine days after the earthquake. It is apparent that in achieving the re-opening what took place is a model of organised and effective response. In dealing with an emergency on a railway, experience is vital. For a natural hazard which may occur once in a lifetime, it is equivalent experience that is used to tackle the key issues. For a railway the most equivalent experience to a natural hazard is a derailment.

In attempting to analyse the Edgecumbe earthquake and set out recommendations for effective response, reference is also made to the experience gained by the author in dealing with derailments.

Response to exceptional natural hazard

The first priority, whilst the emergency services are dealing with the casualties, is to establish whether any assistance can be given in the evacuation of the injured and fit to a place of safety, and, in the period until they are able to fend for themselves, to ensure that there is adequate sustenance for their continuing state of health. In remote areas of harsh terrain the railway could be the principal link between centres of civilisation and the railway maintenance system may be uniquely placed and must be ready, especially in undeveloped countries, to take control and manage the emergency situation.

In a derailment in a remote part of the Sahara some 30 years ago, passengers perished for a lack of drinking water while the railway authorities were unable, by usual means, to provide it. Control of the situation was assumed by headquarters when it would have been preferable to give authority for local initiative. In an abnormal situation the risk to life must be assessed and then recourse made to any available means to safeguard it. The armed services, coastguard, road and air services can all assist to further protect life in a tragedy.

National and local radio reports are a source of early information. However they may be inaccurate and exaggerated and there is no substitute for accurate situation reports from experienced people. As well as the emergency patrols, specialists should be despatched when it has been established how it is best to travel. A recourse to air observation may have to be considered.

A centre of operations must be established together with a reporting-in procedure. The site observers should be equipped with means to communicate with the centre. VHF radio may be used if the network is still operational but if not, UHF radio should be used with a communication chain.

From the site reports the person in charge compiles a situation and damage summary. This report then provides the basis on which immediate action and repairs are planned. Ideally this report should be ready as soon as possible but the target for completion should be not more than 24 hours.

The summary must identify the trains that are trapped in section and where, and what passengers and freight are stranded. If not already inhand the immediate action must be for the evacuation of passengers to a place of safety. Perishable freight should be moved out by any available means, preferably to the customer or temporarily stored in suitable conditions. If neither of these options is practicable the produce may have to be disposed of, but not dumped if it can be used otherwise in the emergency.

The plan of priority repairs should provide for a progressive approach towards the stranded trains allowing for the recovery of the freight and rolling stock. While repairs are proceeding, alternative services, both passenger and freight, should be considered to meet the needs of customers.

Finally, after repair of tracks and structures, maintenance of the restored infrastructure must be included for a progressive improvement up to the complete restoration of normal working. The plan must anticipate the resources required, the personnel and materials needed and available, the means of delivery and the route to site. Wherever possible self-powered equipment should be supplied with sufficient reserves of fuel, supplemented by lighting equipment for round-the-clock working and hand tools.

The availability of standard components is usually good but for special items such as turnouts it is advisable to plan for contingencies with stores of materials which for emergencies need only be second hand.

For round-the-clock working, food and beverages will be necessary to sustain a cheerful work-force and also a system of communication to the families at home to reassure both the workers and the families.

Design and response to the natural hazards of weather
Storms and flooding

It must be the aim of the railway engineer to reduce direct flood damage to the railway. The cost and inconvenience of diverting trains, or terminating them and providing alternative arrangements, of carrying out remedial works often in unpleasant and difficult circumstances should be unacceptable on a year to year basis. Where flooding and damage may be expected most years, policies of long-term prevention are necessary.

Storms cause flooding which causes scour which can damage

(a) bridge footings
(b) embankment toes
(c) stone protection or piling
(d) cesses and shoulders of railway track
(e) training walls and river banks.

Remedies are

(a) gabions and/or mattresses
(b) additional piling
(c) diverting water courses
(d) strengthening training walls and river-banks.

Prevention of damage by design can include analysis of the location of flood damage and the interrelation with the track. In Sudan it was identified that most problems occurred in a few places where the terrain was flat and it was clear that when laid new in the 1950s the track followed the easiest alignment and profile across flat ground and did not consider the possibility of flooding and its consequences.

As some of the track was due for renewal it was suggested that this track be moved to a new alignment on higher ground where the stream beds were better defined and the course of flood water was more predictable. This policy would be more expensive in the short term because it would involve new bridges to cross the water course but as the volume of water would be less and its path could be controlled the problems of washouts and destruction of bridges and embankments would not reoccur leading to substantial savings in the long term and a more reliable service of trains in adverse weather conditions.

Washouts-Sudan experience. The rainy season in the west of Sudan extends from late June to the end of September. During this period, severe storms of short duration but providing heavy rainfall may be experienced.

Absorption of the water by the ground is inadequate and a rapid run-off is experienced with water collecting and flowing in dry stream beds and ground depressions. The soil is mostly fine grained silt and sand in varying proportions and cohesion is negligible. Underlying soil quickly becomes muddy and impervious when wet and it is also unable to resist the scouring effect of running water.

The location of a stream bed can vary from year to year due to the movement of blown sand and the accumulation of debris, especially in the three months prior to the rains. The volume of water in the old stream beds and the transported debris cause erosion of the banks and this can move the position of the stream. The occurrence of the storms is random as also is their location.

Structures such as bridges and culverts constructed to transfer flood water through the railway embankments following one year's inundation may be in the wrong place for floods in succeeding years.

Railway embankments are built from the surrounding soil and compacted under traffic. Rigid structures built within the embankments are liable to be weakened and even destroyed by scour of the foundations, the protective walling and abutments. Displacement of bridge piers, scour of bridge abutments and foundation rafts sometimes leads to total collapse of the bridges.

There is frequent failure, caused by scour, of culvert head walls and embankment protection constructed in cemented stone pitching.

Recommendations. Where diversion of the track was not practicable then strengthening of the structure had to be considered. This took the form of protecting critical structures such as bridges, retaining walls and culverts by sand bags and gabions. In both cases as scour at the base of the structure occurs, settlement of the sand bags or gabions takes place so that protection is continuously afforded.

Gabions were recommended for bridges at the approach including a distance on either side of the embankment of 20 m to protect against damage from stream debris, in the stream bed under the bridge and at the outlet to resist the scour from turbulence and similarly for major rows of pipe culverts and embankments where the stream is expected to run parallel to the railway.

The possibility of constructing new embankments in tetrapods was considered. These would not only be strong to resist the initial impact of the water but would be impervious allowing water through the embankment whilst also resisting scour. This idea seemed too radical at the time and was not taken up, traditional methods of importing dry soil to reinstate the embankments being preferred.

Any solution should consider the availability of local materials and use these for embankment reinstatement. In the case of Sudan, none of the more usual materials — ballast, quarry waste, or broken concrete sleepers were available.

Sand and snow

Although usually occurring at opposite extremes of temperature, the behaviour and effects of sand and snow are nevertheless similar in many respects. Particles are carried in suspension by the wind, which is often strong enough to carry the particles around obstacles. On encountering suitable topographical features the particles may be precipitated in considerable quantities. The actual quantities of material being transported may be quite considerable and a quantity of many hundreds of tonnes per square kilometre would not be exceptional.

Because of its length, the railway installation most at risk from deposition of wind blown particles is the track. Wind- blown sand and snow accumulates in several ways but of interest to the trackman are those accumulations due to obstructions. The 'wind shadow' of an obstruction is filled with vortices of the air flow, of which the average velocity is less than the main airstream, and thus much of the airborne material is deposited adjacent to the obstruction. The severity of the problem is dependent on the type of obstruction. Considerable maintenance costs can be incurred if frequent removal of material is required.

Possible control methods.

(a) Prevention, assisting particle movement to avoid deposition over the critical area. Aerodynamic profiling of cuttings and embankments and the immediate track environment will reduce deposition by allowing a more streamlined airflow. The windward edge of a cutting creates eddies which could cause deposition from a particle-laden wind. Narrow steep sided cuttings suffer most. Where cuttings are wider with flatter side slopes the problem is reduced, not only because of the more streamlined airflow but also because there is more space within the cutting to accept the deposits. Drifting on to embankments is not usually serious unless obstacles have been inadvisedly placed in the path of the wind. However some deposition will occur if again the side slopes are too steep. In critical areas it is recommended that the side slopes of cuttings should be no more than 1:6 (16.7%). These flat slopes would also encourage the growth of vegetation (see (c)).The location of critical track features such as the switches of turnouts should be away from remote areas. When this is unavoidable the

method of operating the train service should be examined so that when deposition of particles may be expected, the switches can be retained in one position. The accumulation of deposits between the switch and stock rail will eventually prevent the points from closing properly leading to suspension of the train service. The only other sure solution to this problem is to have staff present on site continuously cleaning the points. In remote areas during sand or snow storms it can be dangerous as well as very unpleasant for staff to remain on site to carry out this duty. By providing a relatively hard and smooth surface the migration of particles across the surface is assisted. Paving on the windward side of the critical area can reduce deposition as well as reducing the pick up of particles on the approach to the track.

Paving for track is expensive and could prove impractical as inevitably there must be troughs in the paving for the rails. However oiling of sand by emulsion of petroleum resins has been found to be an effective and relatively economical method for areas of sand deposition.

(b) *Removal of deposits*. This involves the straightforward removal of the material from or nearby the affected area. It is effective but only in the short term. Special equipment such as ploughs may be necessary which, because of the often seasonal nature of the problem, has limited use and is therefore an expensive investment.

(c) *Collection or impounding of material before it reaches the protected area.* Creating an artificial feature on the windward side of the critical area such as a trench may be used to accumulate particles before they reach the critical area. However an equilibrium profile will eventually be obtained at which point, unless it is redug, the trench is no longer effective. This method is relatively short term and requires inspection and maintenance. Fencing is more satisfactory, and open fencing is better than solid fencing. In implementing this system the use of flexible and lightweight materials is an advantage so that the fences can be stored and transported easily. Cheap and expendable local materials are best. For initial installation the base of the fence should be above ground by about 300 mm.

Impounding fences eventually fill solid with deposited material which becomes smooth and streamlined, no longer acting as a trap. The only solution is to successively raise the fence to its practical limits.

Fences may be used in a single row or, where space is available, in a number of parallel rows. In either case considerable land take is necessary if the method is to be effective.

Planting of appropriate vegetation must be considered the best long term solution to the problems of deposition of wind-blown particles. Its main advantages are permanence and visual acceptability. It is probably the most widely used method of control used world-wide. Although relatively expensive in the short term, its long-term benefits in facilitating service reliability and offsetting emergency and maintenance costs should more than provide for its initial cost. The extent of planting will extend well beyond the limits of land necessary for the railway alone. Either additional land should be obtained or permission obtained from the land owner. The programme of planting should be treated as long term, with the sections most at risk treated first.

The provision of a tree line or hedge will supplant the need for a boundary fence. Agronomists should be consulted on the best vegetation to use but in principle it should require the minimum of attention to become established, and not require attention after establishment. Vegetative cover is provided either by seed sowing or by planting established vegetation. Benefits are obtained from planting grasses as well as shrubs and trees. The method and type of plants used depend on the local conditions.

As an example of successful treatment to control sand deposition, the Sudan Forestry Department planted seedlings of mesquite alongside the tarmac all weather road in some sections between Port Sudan and Khartoum, on the southern fringes of the Sahara desert. Within 3 years many trees flourished. The ability of the roots of plants to bind sand is well known but the conditions must be suitable for them to become established. The success of mesquite was because

(i) it requires very little water, just a few days of rain each year is sufficient

(ii) seeds are readily available locally

(iii) when young, the plants are not eaten by local livestock

(iv) when established, goats and camels eat the crop of the tree, a seed pod, and the seeds are then spread around the area of vegetation in the droppings of the animals, thus providing further propagation of the mesquite and extending the area of stabilisation.

In 1987 Sudan Railways Corporation were considering adoption of this technique alongside critical areas of the railway.

Other experiences in Sudan are illustrative of the problems associated with wind-blown sand. Problems were caused in areas of sandy soil by loose sand blowing onto to the track and rail head. Eventually it is possible for the depth to become such that derailment is possible. In remote parts of the desert the railway may be the only contact between towns and villages and it is natural for the railway to be used as the direction to follow. Frequently loose sand is kicked onto the rails by people and animals using the railway as a walking route. Such sand on the rails creates a noisy ride by rail vehicle and encourages the train drivers to reduce speed. Lorries travel alongside the railway and sometimes approach too closely and push sand onto the rails. In Sudan the track is laid on the native soil. When this is loose dry sand it is easily vibrated out of position, and at rail joints in particular low joints and misalignments are common.

Recommendations were

(i) to maintain rail level at least 400 mm above surrounding ground so as to discourage trespass, to import more stable soil with a silt content which allowed better compaction and provided a profile more resistant to wind erosion, and to reduce the upper surface of the containing soil to the top of the sleeper so that a reservoir was provided to accept some wind- blown sand without reaching rail level

(ii) where large sand dunes were present alongside the railway these were to be bulldozed and the sand removed to a place remote from the railway downwind

(iii) in very dry areas which suffered from the 'haboob', a hot dry wind blowing north-south across a track alignment mainly east-west, to have staff and sand ploughs on standby to clear accumulations of sand

(iv) in areas which experienced a minimal annual rainfall to plant vegetation, in particular mesquite (see above).

Conclusion

For the railway, natural hazards are a fact of life. Whilst they cannot be avoided, their effects can be minimised by an efficiently managed response, and where economical to do so, by design.

It is the Author's opinion that local knowledge and experience must be mobilised to play a significant role in both response and design. The railway maintenance system is well used to dealing with both large and minor crises and has much to offer in minimising the effects of natural hazards.

References

1. T. J. RAWLINGS. The Edgecumbe Earthquake. *J. Permanent Way Institution.Part 2.* 1989
2. D. G. SPERRING. *Civil engineering requirements.* Sudan Railways Report commissioned from Transmark by the Overseas Development Administration. November 1985.

Discussion on Papers 6 and 7

H. M. AHMED, former General Manager, Sudan Railways

I would like to highlight some points relating to D. Sperring's Paper.

The accident at the Sahara (mentioned by D. Sperring) was very true. But regarding the arrangements for rescue of passengers and repair of track etc., it became a practice that in case of derailments, accidents, hazards etc., it is the duty of the Governor and his staff in the nearest area to take care of the matter (provision of food, water, first aid, transport to hospital etc.). Other people also contribute.

On certain occasions we have made use of the armed forces, police force, fire brigade etc. This is particularly common in the Southern province where there is civil war between North and South Sudan. When I was shot during work in the Southern Line extension, it was the police force at the nearest main station that came to my rescue.

There are certain hazards which can be expected. Rains are generally expected because they happen mainly during the rainy season (July/September). Only the volume of the rain or washout of track is not known. Therefore preparations for washouts should be adhered to, tools, food and equipment for flood fighting should be ready for an emergency.

Washout experience in Sudan

People have become expert and are very quick in repairing the damage done by washouts to open the line quickly for trains. The track runs on an earth embankment; this is why it is easily affected by flood. Also there was little knowledge about the topography of the land and because streams keep changing their courses, bridges have been constructed according to past experience of places damaged. There are now a lot of idle spans or bridges.

Recommendations mentioned

For strengthening the bank where diversion is not possible measures have been applied. The latest bank protection was implemented by an Italian company (Rechii) in the western extension against an Italian Grant to Sudan Railways.

Availability of local materials

Stones, ballast and quarry waste are available but no concrete sleepers.

DISCUSSION

Sand and snow

In Sudan we have a problem of sand which causes a lot of hazards to the track which can lead to accidents and derailments. Sand ploughs are now used. A green belt could be a solution using mesquite plants because it does not require a lot of water. But still we need a good proposal for the places where we have a lot of moving sand and don't have water at all for planting trees. We also have the problem of exposing clay soil which causes a lot of track irregularities during the rainy season.

8. Manual methods– the alternatives of length gangs and mobile gangs

Professor J. K. MUSUVA, Executive Chairman, Kenya Railways

History of Kenya Railways

Kenya - Uganda Railways was built by the British Government as a means to stop the slave trade and also to tap the hinterland produce. The construction was started in 1896 from Mombasa with the original intention of reaching Victoria Nyanza. The expenditure for the construction was availed by the British Government. The engineering survey work started during 1895 but construction did not become a reality until 1896 when Mr George Whitehouse, Chief Engineer, arrived at Mombasa. He was accompanied by a team of engineers, surveyors, masons, carpenters and labourers from India where he had worked before his new assignment.

Track laying and surveying works started on arrival of the advance team with a lot of difficulties encountered en route. Most of these difficulties had been expected since a route survey had not been undertaken. The land terrain from Mombasa to Nairobi was easy all through until the construction reached the Rift Valley Escarpment where a steep drop down the valley had to be constructed. Rope inclines had to be used for dropping or carrying the wagons into or out of Rift Valley until 1900 when a permanent line down the escarpment was built in 1901. A better realignment has since been constructed.

Before reaching Nairobi, the construction progress was interferred with frequently by the lions better known as the "man-eaters of Tsavo" at that time. Some of the coolies — Indian labourers — and engineers were killed and eaten by these lions. Diseases of various description also delayed the progress of the work greatly since most of labour force were not used to mosquito bites.

In 1901, the railway line eventually reached Port Florence, (now known as Kisumu) on the shores of Lake Victoria after five and a half years, covering a distance of 580 miles. On reaching Kisumu, it was decided to extend the railway line further in order to connect with Uganda via the lake services. The total construction cost up to the lake shore was estimated at £5.3 million.

Initially the Uganda Railway permanent way consisted of flat bottom 50lb steel rails laid on steel troughs and timber sleepers. The construction of railway line has no doubt been the most important event in the history of

East Africa. It has transformed the hinterland and practically created all the towns which exist in Kenya and East Africa at large.

When the railhead reached the shores of Lake Victoria in 1901, the settlers who had arrived during the same period decided to have a rail link to Uganda. This was later approved after a protracted dispute regarding the route and the funding facilities. This extension was then finally built as a branch to Uganda from Nakuru but it was later upgraded to main line since it was a link further to the West into Congo (Zaire). This extension was implemented between 1924 - 1928.

During the same period, a few branch lines in Kenya were also under consideration for construction. A link to Tanganyika (Tanzania) — the Voi-Kahe line — was constructed at the same time.

The then Uganda Railways (UR) gave birth to Kenya and Uganda Railways which later, during the first half of this century, amalgamated with Kenya Harbours to form Kenya and Uganda Railways and Harbours (KUR&H).

The two Railways in the East Africa Region (Tanganyika Railways and KUR&H) which had developed independently of each other later amalgamated to form the EAR and Harbours under the defunct EA community. At the break of the Community the three EA states resorted to operating rail and lake services on their own within their boundaries.

Track modernization

Kenya/Uganda railways was constructed using manual labour; this was basically because there was no earth moving machinery locally available and the labour was cheap, imported from India, though towards the end of the construction some of the local people had also joined the project.

Track maintenance has virtually remained manual since the time of the construction. The track was laid on barely prepared formation and good soil from the surrounding area was used for packing the track. The line did not carry much traffic other than being used by the local settler farmers to export their farm produce. The track was laid in light permanent way materials, 50 lb, from Mombasa to the shores of the Lake Victoria. Traffic density and the axle loads increased progressively over the years necessitating the upgrading of some of the lines. The Mombasa-Nairobi section has consequently been re-laid in 95 lb track materials, Nairobi-Nakuru/Malaba in 80 lb materials and Nakuru-Kisumu in 60 lb materials. The Nakuru-Kisumu section is currently being re-laid in 80 lb track materials as the traffic density on this section has increased significantly due to generated traffic from and to Uganda via the lake route and consequent rail failure as a result of fatigue and old age. Following the need to upgrade progressively some of the lines and provide further rail services, additional branch lines e.g. Rongai/Solai,

Gilgil/Nyahururu (Thomson Falls), Nairobi/Nanyuki, Liseru/Kitale and Kisumu/Butere Branch lines were subsequently constructed using released 50 lb permanent way materials. Following increased traffic density and axle loads as a result of generated industrial activity at Thika Town, a section of Nairobi/Nanyuki branch line, viz, Nairobi/Thika was recently in 80lb permanent way materials.

Methods of track maintenance in Kenya Railways
Manual track maintenance

The two main methods of manual track maintenance in practice in the Kenya Railways system are the Orthodox and the Flying Gang Trolley methods. The former method is as old as Kenya Railways itself and the latter was evolved following the introduction of light motorized inspection and material conveying trollies into the system and the need to provide welfare facilities to members of our maintenance staff who were under the orthodox maintenance system stationed in remote and poorly served areas. The Orthodox method is used on both the trunk and branch lines except the Magadi Branch line. The Orthodox method is a system whereby a maintenance unit (gang) are accommodated alongside the railway line and each gang covers a track length of about 7 km depending on the standard of track and formation conditions. The basic method for calculating the strength of a unit depends on the number of curves, lines and turn-outs etc., and is worked out in accordance with Table 23.09B of our *Engineering Manual :Vol.1. Technical Instructions*, of 1962 - see Appendix III. The Flying Gang Trolley maintenance method basically involves the use of mobile gangs who are conveniently housed at depot stations and are carried to the site of work by motor trollies in strict compliance with our trains' operating procedures. The size of this gang is larger than one lineside gang and works under the supervision of a Sub-Inspector on a daily basis. The system was considered beneficial in a social\economical\welfare sense and was introduced on the majority of the branch lines on the defunct East African Railways and Harbours network. In Kenya Railways, the Magadi Soda Branch Line is still being maintained under this system. This type of maintenance is adequate for branch lines with relatively low traffic where high speed is not a consideration and difficulties such as unstable formation etc. are of a minor nature. A typical manual maintenance programme for section gangs on stone ballasted track practised in Kenya Railways has been designed to cover the whole calendar year. This is depicted on Annexure 29 of our *Engineering Manual: Vol.1* (see Appendix I). The programme is well balanced to take care of wet and dry months of the year and has over the years proved suitable especially with orthodox method of track maintenance. The success of this programme on the Flying Gang maintenance method has been dictated by availability of motorized trollies for staff and material deployment and procurement of successful working line possession on our single line operation. For earth\murram ballasted track the activities are re-organized to accommodate through packing when the moisture content in the ground is optimum and therefore practical to undertake packing, clearing and cutting drains and waterways just before the beginning of heavy rains.

Mechanized track maintenance

Three tamping and lining machines were acquired during 1985. This lot was preceded by a mechanical tamper which had served for a long time. The success of these machines depends on the daily availability and procurement of successful working line possessions on our single line operation. As a result the machines have been supplementary to our existing maintenance programme rather than a replacement of our manual maintenance staff or gradual introduction of mechanized track maintenance. Procurement of spare parts from overseas to sustain availability has been our main constraint.We have hitherto depended on Thermit welding of rail joints the quality of which has been difficult to maintain in the field. Consequently we have experienced cases of welded rail joint failure. As a solution to this problem we have recently acquired a mobile flush butt welding machine which has already been commissioned and is currently busy on the on going Nakuru/Kisumu upgrading and rehabilitation works. Once again the output of the machine is determined by the daily availability and procurement of successful working line possession in our single line operation.

Track maintenance standards

An automatic track fault recording car acquired during 1985 has never produced any results because of a permanent breakdown since it was commissioned. This machine was preceded by an obsolete mechanical one which had worked satisfactorily for a long time. Although track maintenance has been undertaken to the specification and entire satisfaction of the Civil Engineering Department it has not been possible to compare maintenance standards of the various section for a long time.

District organization - Civil Engineering Department

Kenya Railways comprises three Districts - Lake Engineering District, Nairobi Engineering District and Coast Engineering District. Each District is headed by a District Civil Engineer assisted by three Engineers (see Organization Chart, Appendex II). All the Engineers are properly trained on Railway Engineering both locally and abroad. In the District organization, the District Civil Engineer is responsible for the entire district track network assisted by a deputy and two or three section engineers. In the office of the District Civil Engineer, there is a District Permanent Way Inspector who is a very senior permanent way inspector with wide and varied work experience and on track materials. He handles most of the track problems on behalf of the District Civil Engineer as directed. To ensure that closer supervision is implemented, the District permanent way is divided into sections and sub-sections. Each section is headed by a Permanent Way Inspector, (PWI), and under a Permanent Way Inspector there are Permanent Way Sub-inspectors (SPWI) depending on the size of the section under his jurisdiction. Under the SPWI, there are the maintenance gangs and each gang covers a track length of about 7 to 8 km. The PWIs and SPWIs are very well-trained supervisors (Track Inspectors) and their basic training both at the Railway Training Institute, and in the field takes about 3 years and before an SPWI is assigned a sub-section his capability in track maintenance and repairs and personnel management must be ascertained. The same is done for the PWI who must show quality leadership, good knowledge of track repairs and maintenance, good personnel management skills, all round and wide experience and ability to work independently. As regards a maintenance gang, especially in the orthodox track maintenance method, the number of labourers (gangmen) is determined by equivalent length of plain track. As already stated before the equivalent length of plain track is assessed by using conversion factors shown in Table 23.09B of our *Engineering Manual: Vol 1* (see Appendix III). The basic ratings were standardized on the basis of the weight of track, traffic, density and formation conditions as shown in Table 23.09A of our *Engineering Manual Vol.I* (see Appendex IV). Each gang has a leader designated a Ganger (Jemadari) with an assistant designated a Keyman, (Chabiwallah). The Gangers selection is based on his experience, knowledge of work and reliability and ability to lead and work independently. Education has not been a criteria for this appointment but the need to take this into account is on the increase. Seminars are periodically organized for the Gangers where they are taught basic facts about their daily track assissgnments and implementation procedures to be followed in carrying out various track activities . Each Ganger is supposed to exchange duties with his Keyman on Thursdays so that the Ganger walks his entire section inspecting the track and noting areas or sections with faults which he would attend the following week. Exchanging duties gives a basic training to the Keyman so that he can eventually be promoted to or take over from a Ganger.

Manual maintenance programme
Procedures

The manual track maintenance on Kenya Railways is carried out systematically to an annual programme, the purpose of which is to ensure that

(*a*) essential maintenance works are regularly carried out over the year
(*b*) works are carried out in suitable seasons in an organized manner
(*c*) section gangs are fully conversant with the works to be done and do not waste their time between the intervals of inspections by the Supervisors or Engineers.

A lot of the time is devoted to through packing and screening of stone ballast which have to be done in fine and dry weather and when the roadbed is dry respectively.

Manual track maintenance
Tools and equipment
A gang (Platelayers) must always have the following equipment with them to accomplish the manual track maintenance of their section

- Beater – one for each gangman
- Fork, ballast – one for each gangman
- Cant Board – to be kept by the Ganger
- Spirit Level – to be kept by Ganger
- Two fishbolt spanners – A pair each for the Ganger and Keyman
- Box spanners for points and crossings
- Tin of detonators
- Ballast Template
- Full set of hand flags
- T-square
- Track Gauge
- Plantation Hoe – one for each Gangman.

Beater packing
The correct way to pack a sleeper is for two men, standing back to back, to beat each end crosswide using the Standard Track Beater with simultaneous strokes. The portions to be firmly packed are 300 mm on each side of the centre of the rail. Each sleeper is firmly packed at the rail seats.

Thorough packing
Thorough packing is normally carried out systematically from one end of the gang length to the other. Thorough packing consists of the following operations in sequence

- opening up the track
- examinations of rails, sleepers and fastenings
- squaring of sleepers
- slewing track to correct alignment

- gauging
- packing the sleepers
- repacking of joint sleepers
- boxing of ballast section and tidying up

Task work

This is generally encouraged in all cases where it is practicable, e.g. such jobs as

- clearing of culverts and drains
- weeding
- lubrication of fish plates and fishbolts
- loading and offloading materials etc. are all suitable for task work, (i.e. piece-meal work).

Cost-effectiveness of manual maintenance

The manual track maintenance system adopted in Kenya Railways has over the years proved very successful and cheaper than mechanized track maintenance as manual labour has more or less remained cheap and plentiful and the cost of machines has gone out of control. Kenya is among the Developing Countries in the World where shortage of foreign exchange is experienced. It is therefore not easy to invest in the procurement and maintenance of the modern machines for mechanical track maintenance as the capital outlay required in foreign money outstrips the local cost of maintaining the existing and successful manual track maintenance. The viability of the mechanized maintenance system is also dictated by our single line operations, train speeds, traffic density and resultant revenue earnings.The most economical number of hours required to optimize the use of say a tamping machine on our system is six hours and this can only be achieved on a line block on our single line operations. This is very expensive due to disruption of traffic from a global or localized point of view and the resultant loss of revenue. In a manual maintenance system, there is no interference with the operations of trains. In addition to these considerations, the manual method of track maintenance is also favoured by the type of steel trough sleepers which is currently in the track. The steel sleepers are light and can be handled by one person unlike the heavy concrete sleepers. The steel trough sleepers are designed to hold the stone firmly after packing has been done properly and this can remain intact for three to six months if not disturbed. Kenya Railways has recently acquired 3 tampers and 1 modern track recorder through a World Bank loan but for reasons stated earlier in this Paper the tampers are only supplementary to the existing manual track maintenance method.

References

1.HILL M. F. *Permanent Way: Vol. I. The story of the Kenya – Uganda Railway*
2.HILL M. F. *Permanent Way: Vol. II. The story of Tanganyika Railways*
3. EAR & H. *Engineering Manual: Vol.I. Technical Instructions.* 1962

Appendix I. Typical maintenance programme for Section Gangs on stone ballasted track

ITEM OF WORK	QUOTA PER ANNUM	COLD SEASON			WARM WEATHER	SHORT RAINS		HOT SEASON			LONG RAINS		
		JUNE	JULY	AUG	SEPT	OCT	NOV	DEC	JAN	FEB	MARCH	APRIL	MAY
1 THROUGH PACKING AND DRESSING	WHOLE GANG LENGTH	**********************************											
2 OILING CLIP.BOLTS OF STEEL SLEEPERS	HALF GANG LENGTH						********						
3 LUBRICATING FISHPLATES & FISHBOLTS	HALF GANG LENGTH							********					
4 SCREENING BALLAST & MAKING CESS	ONE MILE MINIMUM								*******************				
5 CLEARING, CUTTING DRAINS & WATERWAYS	ALL					***					***		
6 OVERHAULING SWITCHES & CROSSINGS	ALL												********
7 SLACKS & MISCELLANEOUS WORKS	-	<——————————— AS REQUIRED ———————————>											************

NOTE : FOR PRGRAMME ITEMS 1 & 4, THE GANGS SHOULD WORK OUTWARDS FROM THEIR SECTION
LIMITS IN ORDER TO OBTAIN CONTINUOUS LENGTHS OF COMPLETED TRACK : E.G., GANGS
1 & 2 SHOULD WORK AWAY FROM EACH OTHER FROM THEIR JUNCTION AT MILE 4. GANGS 3
& 4 WILL DO LIKEWISE FROM MILE 12 AND SO ON.

Appendix II. Civil Engineering Department permanent way organisation structure

```
RS 3          ──────────────────────────────────┤ C.C.E.   │
                                                 ┊
RS 4          ──────────────────────────────────┤ A.C.C.E. │
                                                 │ (P&W)    │
                                                 ┊
RS 5          ──────────────────────────────────┤ P.E.     │
                                                 │ (P&M)    │
                                                 ┊
              ┌────────────┬──────────────┬──────────────┬──────────────┐
              ┊            ┊              ┊              ┊
RS 6   ───────┤ S.P.E.  │  │ S.P.E. │   │  D C E   │    │  S P E  │
              │ (D.R.S) │  │  (M)   │   │ (3 POSTS)│    │   (P)   │
                                             ┊
RS 7          ──────────────────────────────────┤ D.D.C.E. │
                                                 │ (3 POSTS)│
                                                 ┊
R A           ──────────────────────────────────┤ A.C.E.   │
                                                 │ (5 POSTS)│
                                                 ┊
SNR.EXEC."B"  ──────────────────────────────────┤ D.P.W.I. │
                                                 │ (3 POSTS)│
                                                 ┊
EXEC. I
EXEC. II OR   ──────────────────────────────────┤ P.W.I.s  │
                                                 ┊
RB IV-RB II   ──────────────────────────────────┤ S.P.W.I.s│
                                                 ┊
RC I          ──────────────────────────────────┤ GANGERS  │
                                                 ┊
RC I          ──────────────────────────────────┤ KEYMEN   │
                                                 ┊
RC II - RC I  ──────────────────────────────────┤ GANGMEN  │
```

Appendix III. Permanent way gang strength rating

Basic rating without extra factors

For the men per mile excluding the Ganger and the Keyman, on the different sections of the Railway, see Appendix IV.

Factors to be used in assessing additional men

	Equivalent miles of Plain Track
Normal section gangs	
Plain Track	Actual length
Curvature in Plain Track over 4 degrees sharpness	1000 Lft./Track = 0.05 miles
2nd Loops and Sidings	1,000 Lft./Track = 0.07 miles
Turnouts	Each = 0.04
Exceptionally bad sub-grade extending more than half the gang length	1 additional man to the gang
Station Yard gangs	
Plain track	Each = 0.04
Turnouts	Whole = 0.16
Scissors Crossover	

Note:(*a*) In computing the number of men using the Basic rating, with or without the additional factors, over 0.33 man to be taken as 1 man. (*b*) Plain track is exclusive of the space occupied by a turnout, i.e. it is measured to the end of the stock rail on the one side and to the end of the crossing on the other side.

Example

Normal 4 mile section with basic rating of 2 men per mile; 5 turnouts; 3,000 ft. of sidings; 4,000 ft. of curvature over 4 degrees.

	Equivalent miles
Plain	4.00
5 Turnouts x 0.04 each	0.20
3000 ft. Sidings x 0.07 per 0/00 ft.	0.21
4,000 ft. Curves x 0.05 per 0/00 ft.	0.20
Total	4.61

Therefore, Number of men for the gang = 2 x 4.61 = 9.22 = 9 men

Appendix IV. Basic rating of normal permanent way gangs

Section of railway	Men per mile
Mombasa - Nakuru	2.00
Nakuru - Malaba	1.75
Kisumu Branch	1.50
Nanyuki Branch	1.50
Voi/Taveta Branch (after stone ballasting)	1.25
Magadi Branch	1.25
Nyahururu Branch	1.25
Solai Branch	1.25
Kitale Branch	1.25

Basic rating for Station Yard gangs

Station	Men per mile
Mombasa Port	2.00
Voi	1.50
Nairobi	2.00
Nakuru	1.75
Kisumu	1.75
Eldoret	1.50
For all other Station Yard gangs	1.25

Notes: (*a*) The number of men per mile is exclusive of the Ganger and Keyman; (*b*) Where section gang lengths are in kilometres they must be converted to "miles" by multiplying the former by 5/8 to two decimal places; (*c*) On long sidings where separate gangs are provided, e.g. the Sultan Hamud - Kibini Hill Siging, the basic rating shall be 1.25 men per mile.

9. Maintenance equipment – users

J. SVENDSEN, Regional Engineering Manager (North Europe),
Pandrol International

Introduction

This conference concentrates on lightly used track, including railways in developing countries, for which the maintenance aspect is different from heavily used lines in a more technically sophisticated environment. My own experience stems from maintenance of track both in Scandinavia and in developing contries.

Technical characteristics of lightly used track

In the Western world

Lightly used tracks on a railway in a Western industrialised country are usually outside the main line system of the railway, as lines to remote areas, yard tracks or sidings. These tracks are originally mainly laid as jointed track on wooden sleepers, sometimes in gravel ballast. As wear and tear has made the technical standard unacceptable from a train safety point of view, the lines are relaid using second-hand material released when the main lines have been renewed. On most railways no re-laying of concrete sleeper main lines has taken place yet, except on lines with concrete sleepers from a very early date where the technical standard of the sleeper or rail fastening proved to be inadequate, in which case the sleepers mostly have been writen off and scrapped. In many cases the wooden sleeper standard has therefore been retained, but the rails have been welded to longer lengths to reduce the number of joints.

On some lines there is passenger traffic requring a high train derailment safety as well as comfort standard, in which case the line may have been relaid continuously welded on concrete sleepers, also because a wooden sleeper complete with fastenings is sometimes more expensive than a concrete sleeper.

In developing countries

Looking at the tonnage and number of trains in developing countries, even the main line tracks must mostly be characterised as lightly used. Steel sleeper track on gravel or stone ballast was very commonly used, as also were wooden sleepers.

When renewing track nowadays, the trend is going towards use of concrete sleepers. There are a number of reasons for this, mainly

- shortage of foreign currency for import of steel sleepers
- non-availability of indigenous wood
- wooden sleepers susceptibility to rot and termites
- domestic prodcution of concrete sleepers can take place by use of mainly indigenous materials.

When introducing concrete sleepers, the rails have to be continuously welded. Tazaras experiences with jointed track on concrete sleepers clearly demonstrates that. Also the need for a better quality of the stone ballast and for an adequate ballast section is more pronounced with concrete sleepers.

The need for mechanised equipment

In industrialised countries the unit cost of maintenance using manual labour is generally higher compared to the cost using mechanised equipment with similar production capacity. Regardless of the track standard one therefore wants to mechanise the maintenance operations to a higher degree. Also the availability of older machines which are no longer suitable for main line operations, but still can do a useful job on lines where the operating time constraints are not so severe makes mechanisation easily achievable.

On railways in developing countries the problem is more complex. Local labour paid in local currency is competing with mechanised equipment imported and paid with scarce foreign currency.

In the foregoing is elaborated the development leading to the introduction of continuously welded concrete sleeper track. An implicaion of this is the need to use mechanized tamping and lining equipment.

The weight of the track and the lateral forces in continuously welded track makes it difficult to achieve the required maintenance standards working with hand jacks and pickaxes. Moreover, the concrete sleepers are high, which makes it difficult to access the low edge of them with pickaxes, the result being that the edges are damaged by the pickaxes.

Although ballast handling work is more suitable for manual labour it may still be desirable to mechanise it. Ballast cleaning and restoring the ballast profile requires a substantial number of men, which may cause problems for the railway simply in providing housing, food supply, transport, etc. when working in remote areas.

Another area where mechanization is desired is transport and handling of heavy items, such as rails, concrete sleepers, turnout parts, welding equipment, etc.. For this purpose some railways have acquired heavy duty motor trolleys with loading platform and crane. These trolleys are also able to pull a couple of wagons i.e. for ballast transport.

Economy

The technical development on the railways in the industrialised world makes it self-evident that track maintenance work will be most economically carried out using mechanised methods also on tracks with light traffic.

In 1989-90 Kampsax International A/S of Denmark carried out a survey for the Southern Africa Transport and Communications Commision (SATCC) on Mechanised Preventive Track Maintenance including a comparison of track maintenance costs using manual and mechanised methods, taking into consideration

(a) workers salaries
(b) supervision
(c) accommodation and transport
(d) depreciation of equipment
(e) fuel
(f) spare parts
(g) maintenance of equipment.

For the eight countries involved, only in one country where salaries are very low compared to the rest of the countries, was track tamping and lining less costly using manual methods. For two countries reliable cost data were unobtainable.

Concluding reflections

Given that the technical standard of the tracks in developing countries involving concrete sleepers and continuously welded rails is to be retained and extended, there is little doubt that some mechanisation of the track maintenance operations is required.

How sophisticated the equipment needs to be is a matter open for discussion. From the suppliers' point of view, the volume of the market, however, is limited and reliable, as the financial constraints overrule all good technical arguments.

The financial constraints also affect the possibility to maintain the quipment when spare parts supplied with the equipment are used up and new ones need to be imported.

The equipment maintenance is usually carried out by the workshops of the Mechanical Engineering Department and it is experienced that civil engineering plant is given low priority, and that the workshop staff's competence on this kind of equipment is random.

From the user's perspective, the ideal equipment should have the following properties

• low purchasing price

- simple to operate without specialised skills
- high operational reliability
- simple to maintain and repair
- few and standardised spare parts.

The above is obviously true also in a more sophisticated environment, but in particular it is true in developing countries.

10. Maintenance equipment – suppliers

J. A. LEYLAND, Technical Representative, The Permanent Way
Equipment Company Ltd, Nottingham

Introduction

'Enthusiasm without knowledge is not good' (Proverbs 19, v12)

Without knowledge of what equipment is available and what duties that
equipment can perform and without training, no amount of planning and
designing of equipment or methods of work to reduce costs will be effective.

Equally, the manufacturer and the civil engineer have to understand the
needs of the actual user if subsequent designs and the respective technical
specifications are to be effective. Regarding the actual applications of the
equipment, the manufacturer and the civil engineer have common criteria if
they each are to be cost-effective. Regular contact and monitoring of job
needs and equipment capability are essential if planned changes, no matter
how enthusiastically implemented, are to be made successful. These criteria,
for both manufacturer and civil engineer, could be summarised as follows

(*a*) What are the needs of the job?
(*b*) What equipment is available to meet these needs?
(*c*) What available equipment could be modified to meet these needs
without compromising the original function?

Only by regular consideration of these criteria can the use of any plant or
tools be effective in meeting the needs of changing practices and personnel.

Design consideration

In addition to meeting the needs of the job and regular maintenance, plant
must be suited for its task, 'fitness for purpose'. The designer must always
bear in mind the conditions under which the equipment is required to
perform. He needs to understand the application of the equipment, its
working environment and the relationship of the equipment to other pieces
of plant which may be needed to complete the overall task. Railway track
sites can be in remote locations and work is often undertaken at night.
Therefore careful thought must be given as to how the piece of plant can be

got into (and off) site; how it can be cleared in the case of breakdown, how it can be serviced and refuelled, how its working area can be lit.

Permanent way designs and standards are constantly changing to meet changing traffic requirements and mechanical equipment must develop and change accordingly. It must retain a degree of flexibility to be able to cope with a variety of conditions.

In meeting these changing demands, a full understanding of existing equipment is of paramount importance in the event that existing designs can be utilised, there by avoiding unnecessary redesign costs. Indeed such action would contribute to greater plant utilisation, reducing costs further.

Equipment choice

In considering aspects of the relative benefits between simple or sophisticated machines/tools, the point of view is taken by the writer that ideally the civil engineer desires to be independent. In other words, equipment is required that allows work to be done as and when the engineer wants it to be done and not when others would choose. The less need to rely on other services outside the engineer's authority the better control the engineer will have over the work involved.

This is not necessary a poor reflection on the reliability of other support services, though it is a consideration, rather it acknowledges the fact that other services such as training maintenance and transport have other demands placed upon them. The engineer therefore, would have to fit in with other people's priorities and the consequent inflexibility this would cause, especially in times of emergency.

Possession times need to be used to the full and so consideration is needed in assessing how much of this will become non-productive due to time taken within the possession actually getting to or onto the work site with men and equipment. This would reduce available time for the actual work. Equally equipment that is track-borne may of necessity have to travel some distance to site thereby involving potential delays if encountering normal running traffic.

Support services such as training and maintenance need to be considered especially if none exist. The level of support will obviously be reflected in the degree of sophistication of the equipment and naturally affect the overall running costs.

In essence therefore, all these aspects will determine the relative ease of use of each item of equipment. Consequently in an attempt to give a brief and general overview of each piece of equipment a table is provided covering the following aspects which are broadly in accord with the Author's dealings with British Rail (Table 1).

(a) *Skill* – Number of individual skills needed to operate equipment.

(b) *Personnel* – Number of standard permanent way personnel necessary to actually operate equipment. This does not include specialists, look-outs or allowance for the quantity of equipment needed.

(c) *Support (P)* – Specialist persons required.

(d) *Support (M)* – Level of maintenance required

> 0 - Basic workshop
> 1- Regular servicing, knowledge electrical and hydraulics
> 2- Special maintenance support involving electronics
> 3- Special maintenance support involving hi-tech

(e) *Possession* – Degree of difficulty getting to site outside possession.

> 0 - Man-handleable or rough terrain access
> 1 - Requires road vehicle access, readily off or on-tracked
> 2 - As 1 but not readily off-tracked rail vehicle. Cannot get to site out-side possession.

Table 1. Indication of relative ease of use of equipment.

Equipment	Skill	Personnel	Support (P)	Support (M)	Possession
Turning bar	0	1	0	0	0
Permaskate	0	1	0	0	0
Skateboard	0	2	0	0	0
Scooter	0	1	0	0	0
Trolley	0	2	0	0	1
Ironman	0	2	0	0	1
Joint straightener	1	2	0	1	1
Squarer	1	1	0	1	1
Spacer	1	2	0	1	1
Muscleman	1	1	0	1	1
Ballast packer	1	2	0	2	1
with laser	1	1	2	3	1
BREV	2	2	0	2	1
Personnel carrier	1	2	0	2	1
Tug unit	2	2	0	2	1
on track	2	1	2	3	3
Road/rail with crane					
3.5 T GVW	2	2	0	1	0
11 T GVW	2	2	0	1	0
17 T GVW	2	2	0	2	0
Traccess	1	2	0	2	0

It will be apparent from this table that a number of the portable items of plant require reasonable road access. However, if, for example, a rough terrain road/rail vehicle was used for transport, this would give the equipment easier access to site. Indeed if suitably sized with crew cab, a team of men could be totally self sufficient for, say, a rail joint straightening programme, weed control, spot re-sleepering and track inspection.

Maintenance equipment

What follows is a review of selected items of equipment and tools detailing their use and how, if appropriate, they have been modified to provide cost effective solutions to particular needs.

Traccess

A method of transporting rail vehicles up to 7 tonne GVW by road to or near to the work site. This avoids the obvious uncertainties of travelling through the rail network due to interference from the normal running traffic.

Vehicles are transferred directly to the track from the back of the carrying lorry which can be any flatback vehicle 17 tonne GVW converted with winch and flush fitting track and ramp bars.

The system also provides for a portable stillage for line-side pre-delivery storage or speedy off-tracking to allow for passing traffic. Typical time for off or on-tracking is 5 minutes. Specialist skills are not required and the system is operable by two persons, one of which being the vehicle operator.

Providing the obvious limit in size and capacity of the vehicles being carried is acceptable, the system of transport does improve the civil engineers flexibility at having greater control over the transport operation. Indeed, regular permanent way staff can be used and the transport lorry can be used for other duties.

Turning bar

A simple tool for handling flat bottom rail on the ground by turning it over. The specially designed cut-outs on the turning bar head allow it to locate on the rail foot or head and automatically disengage from the rail when it flips over. Two men can turn a 60 ft rail and several can move CWR.

Permaskate

Figure 1 illustrates an appliance used for many tasks, such as allowing long welded rails to be moved easily for any distance, alongside the track. When turned over, it can be used singly as a small trolley for the transportation of permanent way components, eg: a basket of ballast, or a sack of fastenings. It may also be used in pairs (or more) for the transporting of heavier components. They are fitted with highly efficient sealed bearings,

Fig. 1. Permaskate

Fig. 2. Permaskateboard

Fig. 3. Rail scooter

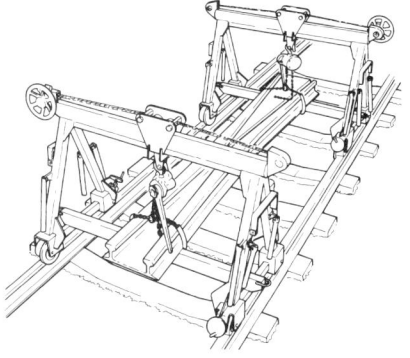

Fig. 4. Ironman

Fig. 5. Rail joint straightener

Fig. 6. Sleeper squarer

Fig. 7. Muscleman

Fig. 8. Ballast packer

Fig. 9. BREV

Fig. 10. Personnel carrier

Fig. 11. Tug unit

Fig. 12. Road/rail vehicle

and weigh only 7.2 kg each, with a load carrying capacity of one tonne per skate. The development of a clip-on top 'Permaskate Board' with detachable handles meant that a more user-friendly carrying surface was available for use with two persons (Fig. 2).

Rail trolleys

These are an essential item of equipment if materials and plant are to be conveyed to site effectively. Since mid 1988, Permaquip have introduced an improved range of trolleys to meet the latest British Rail specifications. Three basic types are available, each complete with a fail-safe braking system and incorporating standard wheel sizes of 255 mm diameter for spares economy.

- Type A – Single piece unit 1220 mm long, SWL 2 tonnes.
- Type B – Split unit 1830 mm long, SWL 2 tonnes.
- Type C – Scaffold trolley, SWL 1.5 tonnes.

Rail scooters

These devices enable rails to be transported along the track, either using the existing gauge rails, or service rails. They can be used in pairs for short lengths of rail (10-20 m), or, with care, in multiples for long welded rail. The capacity of each scooter is 1 tonne. For ease of manhandling, the handle folds, and the unit can be split into two parts, enabling it to be transported in any road or rail vehicle (Fig. 3).

In response to the need for moving singular quantities of concrete sleepers, special attachments have now been made available by Permaquip which can simply fit to any existing or new rail scooter, thereby providing safer working practice with greater plant utilisation. Additionally, these same attachments can be fitted onto carrying bars (8 men required) for moving sleepers when away from the track.

Ironman

A man-handleable mobile lifting gantry fitted with wheels for travel on the track and each having a SWL of 1.5 tonnes. Used in multiples they can be used for cross-tracking and transporting as well as removing old CWR.

In response to a request, a simple refinement allows these units to be used cost-effectively for maintenance installation of half-sets of switches and crossings (Fig. 4).

This single action improved safety because it eliminated dubious manual lifting practices and saved money by reducing manpower and locomotive costs. From a planning point of view, it gave the engineer greater independence and reliability by not having to rely too much on other departmental services outside his control to complete jobs.

Portable rail joint straighteners

These machines are used for straightening dipped fishplated joints. They are capable of exerting a force of 103 tonnes to each joint. This force is sufficient for most rail sections in general use. Dipped fishplate joints, with dips up to 25 mm, can be straightened satisfactorily. Alternative attachments allow the unit to lift by the head of the rail for dipped welds (Fig.5). The majority of dipped welds are in the region of 1-3 mm, although occasionally dips of up to 10 mm are encountered. This machine can cope with these extreme cases. They can be operated singly or in pairs, manually or powered.

The output which can be obtained, with a manually operated appliance, is 10 joints straightened per hour for a single unit, or 20 joints with a double. When power operated (petrol engine driven hydraulics) the output is doubled.

The unit dismantles for assembly on-site. The heaviest component is 117 kg. Time taken to assemble or dismantle the equipment is approximately 2 minutes, and each unit is operated by one person.

Portable sleeper squarer

Out-of-square concrete sleepers were becoming an increasing problem to British Rail with no obvious or cheap solution.

The heavy reliance upon track tamping for track maintenance requires that the sleepers remain more or less square in the track. Further, as the sleeper moves out of square, the nylon insulator is crushed and the track circuit is broken leading to signalling faults. A factor often compounded by the tamping machines.

Permaquip developed a one-man operated portable machine (Fig.6), which could correct the sleeper position without having to open out the ballast cribs or remove the fastenings, a tremendous saving over previous methods which could only be done after ballast removal or picking up and relaying the track. The unit dismantles for assembly on track by two persons, its heaviest component weighing 138 kg.

The development of this unit coincided with a British Rail policy to increase the density of sleeper spacing from 24 to 26, 28 or in some situations, 30 per length. To meet this requirement, the sleeper squaring units were combined to provide a portable sleeper positioner.

Muscleman

This machine (Fig.7) is primarily for use where track lifting and track slewing is being carried out during track removal. It operates on concrete and timber sleepered track, and can be separated into three main sub-assemblies in a few minutes. The main frame weighs 310 kg and has permanent folding lifting handles. These enable it to be carried from the road or rail transport to the working site for assembly. It is manually propelled and a small power pack provides the hydraulic power for the rail clamping, track lifting and track slewing actions. The total weight is 600 kg and it is fitted with a turntable as standard. An effective slew of 350 mm can be achieved per pass, with a lifting force of up to 16 tonnes. Larger slews can be obtained by repeat passes of the machine and fine lining work (by pegs or by measurement from adjacent track) can be obtained in appropriate conditions on unballasted or loosely ballasted track.

With one operator, the average time for an 8 sleeper cycle slew of 300 mm and a lift of 100 mm is approximately two minutes.

Further designs of Muscleman are already in service as motorised units to speed up the work rate, provide quicker access to and from site and reducing manpower needs. This was achieved with the addition of a simple cab arrangement and increasing the engine capacity.

Performance details

Travel speed	Up to 16 km/h
Lift force	16 tonnes
Slew force	8 tonnes
Lift stroke	300 mm (12") plus additional 150 mm (6") of ballast clearance
Slew stroke	200 mm (8")
Effective slew per pass	340 mm (14") Very large slews being achieved by repeat passes of the machine
Performance example	300 mm slew 110 mm lift. Average time per eight sleeper cycle, 2 min.

Ballast packer

This machine (Fig.8) can lift and pack track sufficiently for track speeds of 50 km/h (using sighting boards) or 100 km/h when using laser techniques. It is transportable to site using Traccess and requires one machine operator and one sighting board assistant. These both can be regular permanent way staff though for use with lasers two additional technicians would replace the sighting board assistant.

It was developed for use with track renewal, ballast cleaning and re-ballasting operations. It releases expensive on-track tampers to undertake the more refined application of track remedial work. The net lift capacity of 14.5 tonnes enables the heaviest track, with superimposed ballast, to be lifted. Packing banks on both rail operate simultaneously, or separately, having a packing force of 1.2 tonnes in each of the eight independently powered packing arms. The machine is fitted with a turntable, has dual braking system, and in the event of malfunction it has an emergency recovery system.

141

The travelling speed is up to 22 km/h. When lifting 100 mm and packing every fourth sleeper, it has an output of 420 m per hour.

When used for packing every sleeper, its output is 400 m per hour. The time taken for on or off-tracking is approximately 4 minutes and the machine weighs 3.3 tonnes.

The ballast packer has most recently come into greater use because of its high lift capability. With the advent of 'production line renewals' and use of dynamic track stabilisers to re-open track at line speeds of 200 km/h and 225 km/h, the unit is additionally used to lift the track in advance of the ballast cleaner cutting bar.

Mobile welders workshop (BREV)

The mobile welders workshop (Fig.9) is for providing fast response to emergency situations. The unit will carry up to four 9 m lengths of rail, all welding equipment and consumables, lighting, power for all associated tools, and has an extending canopy for weather protection. It has an integral turntable and can be used with Traccess.

Vehicle data

GVW	7 tonnes
GTW (1:50)	11.5 tonnes
Speed	60 km/h
Speed with trailers	30 km/h

Personnel carrier

The personnel carrier (Fig.10) is designed to meet British Rail need for transporting personnel and equipment quickly to work sites especially in remote or inaccessible locations. Additionally, a high standard quality of interior finish was incorporated, in respect of heating and messing facilities. Options available include; dual driving position, carrying capacity 21 men, towing connections and power take-offs. This vehicle can also make use of Traccess.

Vehicle data

GVW	7 tonnes
GTW (1:50)	11.5 tonnes
Speed	60 km/h
Speed with trailers	30 km/h

A budget range of inspection cars and personnel carriers is also available providing basic but functional transport, some of which are man-handleable off-track.

Tug unit

Versatile flat back vehicle with 4 man cab, hydraulic hoist, hydraulic and electric power take-offs (Fig. 11).

Vehicle data

GVW	7 tonnes
GTW (1:50)	14.5 tonnes
Speed	48 km/h
Speed with trailers	30 km/h
Payload	3 tonnes

The capacity of this vehicle can be increased with the provision of track borne trailers. The tug itself can be used with Traccess.

Larger units are available of the double unit type with twin cabs, cranes, 28 tonne GVW, with a capacity to carry 20 m rail and capability to haul wagons. Such units require designated specialist staff (non permanent way) and full maintenance back-up.

Road/rail vehicles

Road/rail vehicles offer the civil engineer the opportunity to be completely independent in terms of transporting men and materials to site with the minimum intrusion on possession times.

Since vehicles can be driven by road as close as possible to the point of track possession, actual possession time and interference with the normal rail traffic that would otherwise occur in travelling solely by rail are reduced. This is further complemented by the vehicle's on-track capability providing site transport between access points and work site, thereby giving greater use of possession times. The incorporation of a four-wheel drive vehicle can also provide accessibility to inhospitable track sites, overcoming rough terrain and difficult access locations.

Additionally, the road/rail concept provides for greater vehicle utilisation which means lower operating or hire costs.

Being able to combine these road/rail features onto existing production made vehicles, affords further benefits by keeping new design and manufacturing work to a minimum whilst providing a vehicle of proven performance.

This is achieved by the fitting of Fairmont Hy-Rail bolt-on rail guidance wheels that give standard road vehicles a road/rail capability. A range of Hy-Rail units exist to cover vehicle gross weights ranging from 1.3 tonnes to 32 tonnes though for general permanent way use, Permaquip have standardised on 3.5, 11 and 17 tonne GVW vehicles.

The Permaquip range of road/rail vehicles are designed to allow them to be on-tracked at virtually any access point quickly, using either an integral turntable or our unique side shift mechanism. Additionally, there are no specialist skills involved that would not already be familiar to regular permanent way staff. Equally so with maintenance except where certain electronic features might be incorporated. Indeed the capability to convert standard road vehicles means that indigenously produced vehicles can be considered, thereby maintaining familiarity for the maintenance workshops.

Multicar

Figure 12 shows a road/rail vehicle which is 4-wheel drive, 3-way tipper model, payload 1 tonne, road speed 88 km/h and rail speed 60 km/h maximum. The epitome of versatility with some 20 basic units covering; tippers, weed spraying units, crew cabs, turntable ladder and hoists, an ideal small gang vehicle for maintenance work. It can also be equipped with hydraulic power take-off and crane.

Hydraulic hand tools

As part of Fairmont Railway Motors, Permaquip market a range of hydraulic hand tools covering some 70 different varieties. With the possible exception of disc cutters and chain saws, no specialised skill is required to operate them. Hydraulic hand tools are hose connected to either portable power packs or suitable vehicle power take-off points. They are to be compared with a number of engine driven equivalents and pneumatic against which they have the specific advantages of a high power-to-weight ratio. This means not only a more powerful hand tool for the operator, but one which is relatively lighter. There is of course, the question of maintenance and consequent fewer engines to maintain with a set of hydraulic tools. A factor further evident when vehicle power take-off points are used.

Individually, engine-driven units when available do by contrast enjoy a reputation for portability (no hose connections) and it is the Author's view that these are best used in situations of difficult access requiring spasmodic attention.

In other words, when work is of a duration to justify several hours possession with consequent vehicle access, particularly road/rail, the benefit of hydraulic hand tools can be realised in terms of reducing down time for repairs (and theft) and keeping tool maintenance work to a minimum.

Conclusion

In considering the complexity of options open to the civil engineer in terms of plant and consequential effects this could have on training, maintenance and transport within the civil engineer's organisation, road/rail vehicles clearly offer a low risk option with maximum independence.

Road/rail vehicles can have a degree of sophistication to suit the engineer's organisation without limiting the vehicles effectiveness and so keeping extra training or maintenance to a minimum.

Since 'normal' maintenance work does not involve high volumes of staff or materials, it is possible to have a self contained road/rail maintenance teams equipped with hydraulic tools. This cost effective method of work is enhanced by the fact that the more immediate costs of purchase and maintenance are minimised.

- The full cost of the road/rail vehicle is offset by the cost of the conventional road unit it would be replacing.
- Any extra maintenance would be limited to Hy-Rail the road/rail gear, since the vehicle itself would be a familiar item to maintain.

Discussion on Papers 8 – 10

K. J. PITKIN, Consulting Engineer, Plasser and Theurer Serv and Cons, South Africa

South African railways do not own their own tamping machines, they put tamping out to contractors. Availability of machines is now 95% and costs are low. The number of tamping machines used has been reduced greatly because they are only used where work is necessary instead of automatic indiscrimiate use. Mechanical track maintenance is essential.

P. EATON, Stanton Bonna Concrete Ltd, UK

Referring to the subject of twin-block sleepers, I would like to make the following comments about the sleepers produced by ourselves in association with the French company, Sateba.

(*a*) The tie-bar is made of rail steel quality material and is 8 mm thick in section. The subsequent rate of corrosion is very slow with the sleeper having a life expectancy of 40-50 years, similar to monoblock sleepers.

(*b*) We produce a range of sleepers to suit different applications. The weights range from 160 kg to 230 kg. An example is the VAX U20 sleeper.

- weight 160 kg
- max speed 140 km/h
- loading 20000t, 25t axle load.

(*c*) Sateba's experience with twin block sleepers dates from the 1920s with many of the original sleepers still performing satisfactorily in track. Sateba has since established manufacturing units world wide with twin block sleepers being successfully produced and installed-in the following countries: UK, France, Italy, Morrocco, USA, Greece, Pakistan, Mexico, Venezuala, Egypt.

E. WEBBER, London

Mr Svendsen said that concrete sleepered track requires mechanized maintenance because the weight of the sleepers and the packing of them make manual maintenance impractical. No one has, however, made any mention of twin block concrete sleepers. I know they are not universally liked, but there are claims that they can be handled manually. Do the Authors believe that they are a suitable alternative to mono-block sleepers, particularly for railways in developing countries, or do they believe that they are not capable of being maintained by manual methods?

J. SVENDSEN, Author

A track panel of twin block sleepers is still heavy to lift with jacks when fully ballasted. More important, however, is that the access to the area where the tamping action takes place is not better using twin-block instead of monoblock concrete sleepers.

In most cases twin-block sleepers are both higher and wider in the rail seat area than their monoblock counterparts. For maintenance work other than tamping I cannot see significant differences.

11. Planning of the work

A. J. J. LINDSAY, MA, MICE, InterCity Civil Engineer, Preston

This Paper discusses some of the various issues which need to be considered when planning cost-effective maintenance of railway track. The experience of BR is used to describe the processes involved.

Introduction

The civil engineer responsible for maintaining a railway administration's track and structures makes a major contribution to the success of that undertaking.

An organisation may be deemed effective if it is achieving the goals. For the civil engineer to be cost-effective implies that the railway infrastructure is maintained so as it contributes to the organisational goals at minimum overall cost. This can only be achieved by the civil engineer fully understanding the business aims and the role to be played in meeting these aspirations.

The civil engineer consequently cannot plan work on purely engineering criteria but must also take into account effects on the overall business. What may be a cost-effective solution on one type of railway is not necessarily the correct answer for a different railway.

A high failure rate on a high density urban commuter network, for example, would not only have a major adverse effect on the efficiency of that railway but may well also impact heavily on the financial performance of the community served. In this case, relatively high cost maintenance solutions which guarantee reliability may be the cost effective solution. Conversely, a low density rural line is not likely to be particularly sensitive to failures and minimum cost solutions may be more appropriate to ensure the viability of this line.

The revenue generated on many BR InterCity trains is journey time sensitive. Significant improvements to journey times have been made by reducing the numbers of short term speed restrictions imposed after track renewal operations by handing back sites to traffic immediately at line speed. This has been achieved by planning longer track possessions and utilising improved methods of ballast consolidation. Whilst these techniques have produced a small cost saving to the civil engineer, the major impact has been to increase revenue to the business thus improving InterCity's financial position. There are many other examples of the civil engineer's work being tailored to improve the service given to the business whilst not necessarily improving the civil engineering unit costs.

It is clear that the railway civil engineer must be capable of planning work not just to minimise infrastructure costs but also taking into account the wider implications to his employers and their business aims.

For this reason those involved in planning must be able to converse in financial as well as technical terms. Financial awareness needs to be present in all levels within the Civil Engineering Department.

This Paper examines the principles to be applied to planning work on railway routes where track possession times are limited by traffic considerations. The Author's experience of planning work on many types of route within the British Railways network is used as a basis for the paper.

Organisational issues

In order to plan effectively, the civil engineer's organisation must be designed to facilitate this function at all levels. Planning is an integral part of the infrastructure maintenance process and will not succeed if it is viewed as a discrete operation within the organisation.

Total quality management (TQM) principles are now being adopted throughout BR. Fundamental to TQM is empowering staff to take decisions at appropriate levels without slavishly adhering to rank and status within the workplace. Dialogue and close working relationships must be encouraged at all levels both within the engineering function and with other departments involved.

To maintain overall control of the process strict financial systems and documented quality targets need to be in place which specify the boundaries and authority delegated to individuals within the organisation.

It is important to acknowledge that the real experts in the day-to-day planning of many aspects of infrastructure maintenance are the Supervisors and operatives who actually carry out the work. These key staff must be given an input into the planning process and, indeed, must be relied on to manage day-to-day planning issues with minimum managerial input.

Typically, on BR a civil engineer may control two to four permanent way maintenance engineers (PWME), each responsible for the delivery of all aspects of track maintenance on their geographical area. Each PWME's area will be further sub-divided into up to four sections, each the responsibility of a Supervisor. The Supervisor in turn directs a number of maintenance gangs responsible for sections of routes. The man in charge of each of these gangs holds the first line of safety management for his section of track and will arrange for adequate inspections and plan his gang's day-to-day maintenance tasks.

Training is clearly, therefore, a key issue in planning for cost-effective maintenance. The organisation should be designed to identify and address training needs. It must include a strong established training function which is involved in the formulation of the long and medium term plans.

The role of management within this structure is to give strategic direction, to monitor progress and identify where change is needed; to encourage ownership of problems and processes and to create the structure to allow decisions to be taken at the correct level.

The organisational structure must be capable of dealing with various types of planning issues: Long term, strategic plans to determine the ultimate aims of the department, medium term plans which form the basis of the annual plans to determine the ultimate aims of the department, medium term plans which form the basis of the annual programme and are used to compile budgets; short term plans to manage the deployment of resources to meet monthly and weekly programmes and, finally, emergency response to plans to meet unplanned workload requirements.

These various types of planning can often be in conflict and a degree of flexibility needs to be inbuilt to allow for this.

Strategic planning

As stated previously, the civil engineer must be aware of what particular business goals apply to the various routes for which he is responsible. Long term plans reflecting both the business goals and civil engineering functional aspirations need to be in place. The civil engineer will be able to measure success by comparing achievement against these strategies. Only then will appropriate actions be taken to ensure that the department is contributing to a successful industry.

Many diverse factors should be included in a strategic analysis. These may include how the workforce is to be deployed, e.g. in small gangs or in large depots, the anticipation of new technology particularly in the field of on-track plant, the age and life expectancy of track components and predictions of budgetary requirements.

As an example, a track component replacement plan which is currently based on using rail mounted cranes and a relatively large workforce would require major modification if the switch was made to large re-laying trains. Track occupation times, overall manning levels, wagon requirements and even the economics of when to replace sleepers would all change markedly. The civil engineer would have to plan not only for the situation after full implementation but also for the transition period. It would not, for example, make economic sense to continue undertaking major maintenance on rail cranes which are being phased out. During the transition stage, therefore, the reliability of the cranes is likely to be diminished resulting in higher short

151

term costs which would have to be planned for. The civil engineer must foresee this type of scenario and be prepared to address the situation.

Future business forecasts will impact directly on the civil engineer's workload. The loss or gain of a major freight traffic flow would greatly influence component life and maintenance cycles. The civil engineer resource base must be managed to achieve the required change in workload in a controlled manner. A gradual reduction in manpower through natural wastage may well be a cheaper solution than incurring redundancy costs. Similarly a structured, gradual increase in labour would allow for adequate training and assimilation of new recruits into the organisation.

Medium term plans

The specific workload requirement for the forthcoming years needs to be established in advance to ensure that it is encompassed within the business financial plan and that it also corresponds to the civil engineer's resource base.

On BR, major work is proposed for inclusion in the programme by the local Civil Engineering Manager (see Fig.l). These proposed items are then inspected by the civil engineer some three to four years in advance of the programmed year and given a priority based on engineering need and business requirement. The desired programme year is confirmed at these inspections. The proposed work is then reappraised again prior to the programmed year and validated against the budget and the resource base available. This system is managed through a computer programme (Computer Renewal Of Way System) which provides a financial and resource estimate and is used to plan resource allocation.

On many medium and high density routes a critical resource is the availability of track occupation times. The civil engineer and the other technical departments agree with the train operators the periods of track occupation to be set aside for use by the Engineering Departments. These are published and are referred to as the 'rules of the route'. The Engineer then will normally plan his work to correspond to the rules of the route.

The planning of major projects benefits from the use of critical path analysis techniques. This process lends itself to the use of computers and there are many commercial packages available which meet the needs of railway civil engineers. Project management is a well documented field and the usual principles will normally apply to railway civil engineering. Often, these projects involve alterations to existing infrastructure and an early decision is needed on whether traffic is to be suspended, to carry out the works, or the work staged to cause only minor disruption to traffic. This decision is clearly in the hands of the business manager but the engineer must

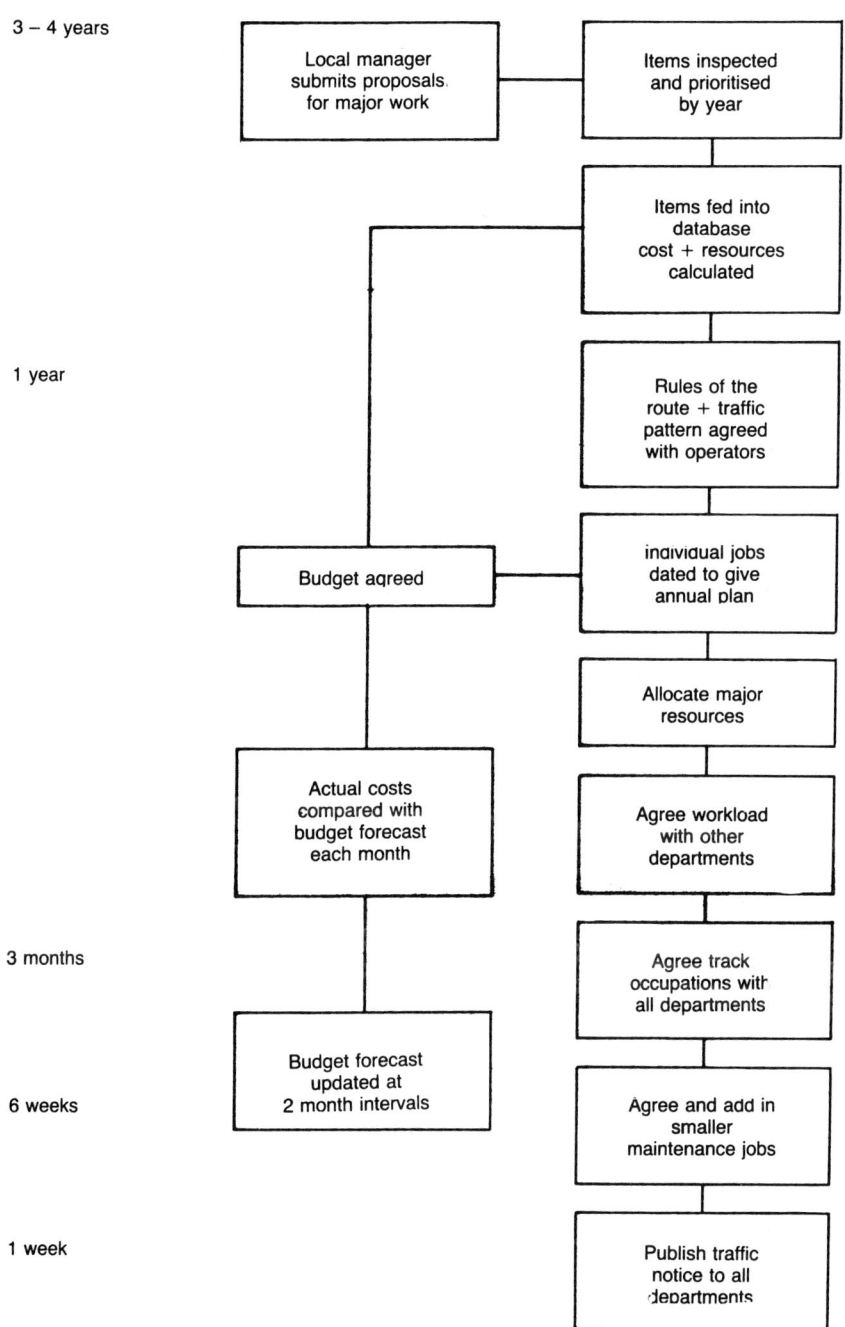

3 – 4 years

1 year

3 months

6 weeks

1 week

Fig. 1. Planning system in use on British Rail

be prepared to furnish comparative estimates to enable a sensible commercial decision to be taken.

When planning and dating the major work for an annual programme, care must be taken to ensure that peaks and troughs in workload are, as far as is possible, kept to a minimum. It would not be economic to plan to meet a resource peak which is only encountered a few times a year. Specific resources such as major plant, wagons and specialist skilled operatives must be allowed for in this analysis. As previously stated, the CROWS computer plan is used to assist in this process.

Short term plans

Once the annual programme has been established and the major work dated, individual jobs will require detailed planning. No matter what the scale of the job involved is, certain elements will always deserve detailed attention.

Safety considerations both for the staff concerned and for rail traffic clearly override all other factors. The safety environment which the Engineer is expected to plan for and provide on BR is increasingly dictated by legislation and it is important to meet the intent as well as the letter of the law. Cognisance must be taken of how staff are to be protected from passing traffic, noise, hazardous substances and dangerous machinery.

It is beneficial that the workforce be encouraged to contribute in the determination of safe methods of work. Too often management attempt to impose working methods without consulting staff with the consequence that the methods may not be practicable and do not have the understanding or support of the staff. It is recommended that all staff be involved in regular (monthly) meetings to discuss safety matters. The feedback from these meetings will be invaluable in developing and improving safe working practices.The safety arrangements for each site should be laid down in detail before work commences and all site staff fully briefed on how these arrangements are to be applied.

The appropriate plant, wagons, materials and skilled operatives must be provided at each site.

Jobs requiring possession of the track should obviously be planned to meet the agreed times of track occupation. For all but the simplest of sites, barcharts showing the start and finish times of the critical processes are an invaluable aid in ensuring this. It is also advisable to indicate on the chart key decision times which represent the points where, if the work is not to schedule, it can be curtailed without transgressing overall possession times. Much of railway civil engineering work is routine in nature and the recording of actual times on the planning bar-charts will enable future plans to be improved in the light of experience. This highlights the advisability of applying the classic process review cycle to planned work (see Fig. 2).

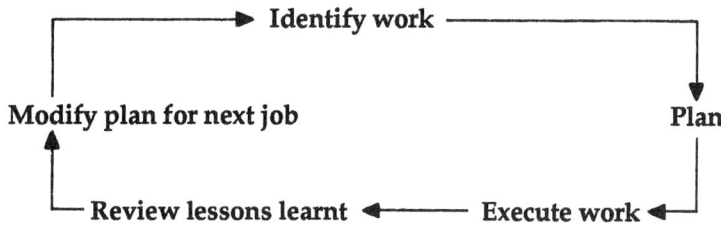

Fig. 2. Process review cycle

Routine maintenance work on BR is normally generated from the local track inspection schedules whereby track is regularly examined at differing frequencies by a Patrolman, the Track Ganger the Section Supervisor and the local PWME. The reports arising from these inspections are used to prioritise work and to compile the weekly work plan for individual track maintenance gangs. A computerised planning system (CAMPS) is in use on BR to facilitate the production of weekly work plans. This programme uses work study in the preparation of the workplan. Workstudy principles can make a major contribution to cost efficiency if used correctly. Those responsible for supplying details for particular jobs (normally done at gang level) must understand why the information is required and how it is to be used. If this is not the case the system is likely to quickly fall into misuse and lose all validity. Used correctly a computerised planning system based on workstudy information can be an invaluable aid to gang planning and can produce historical data from which strategic management decisions can be taken. The CAMPS system is capable of producing work performance details, summaries of how the workforce expended their time and historic records of the work carried out on any particular piece of track.

Emergency response

Emergency response is defined here as any work which falls outside the weekly planning timescales. Such work can arise from many sources and by its nature is likely to be required for immediate operational or safety reasons. Failure to respond adequately to this type of work may involve major cost penalties to the businesses.

Unforeseen and, therefore, unplanned work will disrupt existing plans and clearly is to be avoided if possible. Records of such incidences should be kept and analysed to determine their root causes and attempts made to eradicate them. In many cases trains will be delayed by the occurrence and BR has an automatic train reporting system which logs train delays by cause and location. Examination of causes over a period will indicate where management action can reduce the level of incident. For example, in rural areas animals on the line can often delay trains, particularly during the lambing season. Close scrutiny of train delay data can indicate where it would make economic sense to improve lineside fencing to prevent future delays. Similarly, this data can highlight the failure rate of a track component, such as a particular type of rail joint, which has a significant effect on train reliability.

Nevertheless, unplanned work will always be a factor to allow for. Weekly workplans must be flexible enough to cater for the unforeseen and well established emergency plans must be in place.

Fast response plans depend in the first instance on communications. On BR most routes are equipped with plug-in telephone points at regular intervals and, increasingly, radio pagers and radio telephones are used to provide direct communication with maintenance gangs. Adequate emergency materials should be available to local staff and procedures in place to obtain other resources at short notice. Arrangements must also be established to call out staff outside normal working hours or to provide stand-by staff to cover emergencies. These arrangements must cover all the specialist skills likely to be required.

Budgets and cost controls

Meaningful budgets are financial plans which are accepted by both the provider of the finance and the spender. This implies that the formulation of the budget is a two-way process of iteration whereby both parties engage in dialogue to reach mutually acceptable solutions. This needs to take place at all levels to ensure that everyone who is responsible for delivering against a financial plan is prepared to underwrite and own that plan.

The civil engineer's overall budget should be split into relevant and manageable elements with a clearly defined person responsible for managing each element. These sub-budgets should be structured to give budgetary control to the lowest level possible. On BR the Section Supervisors are involved in budgets for their workload. This encourages ownership of problems and eases the burden on management.

Blanket budget entries such as 'overheads' should be avoided by clearly identifying the specific reason for expenditure and, ideally, the resource incurring the cost. In this example, overheads would perhaps include supervision costs expended against manpower, and accommodation costs expended against manpower (for building maintenance) materials and services (electricity etc.).

Whilst it is practicable to produce separate estimates for each major item of work, much of railway infrastructure maintenance comprises small repetitive tasks. Budgetary provision is made for the latter by the experience of previous years' requirements.

Once the responsible Engineering Manager has compiled his budget by the above process it should be validated against the fixed costs of the known resource base. For example, the summary of many job estimates may not make necessary allowance for the fixed paybill costs of the workforce. This validation process checks that fixed costs are adequately covered and that the various resources are utilised sufficiently.

Once the overall annual budget has been established a similar process should be applied to weekly or monthly periods to produce a spread of expenditure and to again check that the resource allocation and utilisation is realistic.

On BR budgets are set by these methods some six months in advance of the start of the financial year. Once the budget becomes operative expenditure is monitored by job and resource against the budget spread on a four weekly basis and, if necessary, the predicted end of year spend (outturn) amended in light of the emerging costs.

On at least two occasions during the year the financial performance of the overall business is compared with the budget. Individual budgets are then adjusted and new targets set in order to achieve the established financial targets.

Whilst cost control on individual jobs is important and must be encouraged, real large scale savings can only be realised through strict control of the resource base. Reducing the manpower required on a single site will produce little overall savings unless similar reductions are made on many sites sufficient to enable an overall reduction in the number of men employed. Similar considerations apply to other resources such as plant, engine power, road vehicles etc..

Financial control is fundamental to good planning and as costs emerge and are compared with estimated costs the financial data used in compiling future estimates and budgets must be updated.

Attention to financial data will greatly aid cost-effective maintenance.

Conclusion

Cost-effective track maintenance can only be achieved through the correct planning procedures being applied throughout the civil engineering function. In today's increasingly competitive environment the civil engineer must respond to these planning issues in order to provide the service the Railway Industry requires in order to survive and, hopefully, expand.

12. Track rehabilitation in developing countries: direct labour or contractor?

R. MASTERS, Director Mott MacDonald International Ltd., Consulting Engineers, UK

Introduction

The purpose of this Paper is to make comparisons between undertaking major track rehabilitation by direct labour or by contractor in developing countries. It is particularly related to the Author's experience in Mozambique over the last five years.

The criteria by which the two methods are compared are as follows

- quality of completed work
- first cost of rehabilitation
- elapsed time to achieve rehabilitation
- training opportunities
- financial contribution of railway organisation.

Mozambique has three major rail corridors each originally built to serve, in the main, adjacent countries to provide outlets for their exports. The northernmost railway is the Nacala railway linking Malawi at Entre Lagos with the port of Nacala. In the centre of the country is the Beira line linking Beira with central Zimbabwe at the border station of Machipanda, and in the south of the country the port of Maputo is linked to adjacent countries by three railways: to the Republic of South Africa at Ressano Garcia, with the Kingdom of Swaziland at Goba and with southern Zimbabwe at Chicualacuala.

Rehabilitation of the track on all three corridors is either in progress or has been completed. Either direct labour, international contractors or railway companies from adjacent countries have been used. Comparison between the implementation methods will be made where the appropriate data are available.

The projects to be considered are

Limpopo Line: (*a*) Rehabilitation of track between Maputo (km 5) and Chokwe (km 213), a distance of 194 km (there is a 14 km 'break' in kilometerage at km 114). This was carried out by direct labour.

(*b*) Rehabilitation of track between Chokwe (km 213) and km 471, a distance of 258 km. This was carried out by contract between the railway organisations of Mozambique and National Railways of Zimbabwe (NRZ).

Nacala Line: Rehabilitation of track between the Port of Nacala and the town of Cuamba, a distance of 533 km. This was carried out by contractor.

The Limpopo Railway runs from Maputo to the border with Zimbabwe at Chicualacuala. It is some 520 km long and is a single line railway of 3ft 6in (1067 mm) gauge. It is part of the southern system of the Empresa Nacional de Portos e Caminhos de Ferro de Mozambique, E.E. (CFM). Two other lines connect to Maputo port, i.e. the 88 km Ressano Garcia line to the Republic of South Africa and the 69 km Goba line to Swaziland. CFM decided to execute the rehabilitation between Maputo and Chokwe by direct labour and between Chokwe and Chicualacuala by a contract between itself and the National Railways of Zimbabwe (NRZ). This contract was not open to tenders from other contractors but negotiated directly between the two railway organisations.

CFM set up an organisation with the title Brigada de Melhoramentos do Sul (BMS) to direct and supervise the rehabilitation programme for the Southern Railway System (CFM (Sul)). As BMS did not have the skills or resources to undertake this role they received technical assistance (T.A.) in the form of a Project Implementation Team to manage and supervise the Limpopo Line rehabilitation. The Implementation Team consisted of 17 expatriate staff provided by Mott MacDonald under a Technical Assistance Contract funded by ODA of the UK, together with 7 BMS support and counterpart staff. The consultant's Mozambique based activities were supported by a United Kingdom based group who undertook procurement of equipment and materials from the UK and provided general coordination. For the NRZ contract the Implementation Team provided a service which covered the financial administration of contract payments, liaison between BMS and NRZ on materials provision from BMS and advice to BMS on contractual matters. For the direct labour work the Implementation Team provided the supervision and management of all of the activities.

The Nacala railway runs from the port of Nacala to the border station with Malawi at Entre Lagos. This railway is to be rehabilitated in three phases. The first phase from Nacala (km 0) to Nampula (km 192) is complete. The second phase from Nampula to Cuamba is under way.

CFM set up a similar organisation to that set up in the south of the country (BMN) and employed the French Consultants Sofrerail to assist BMN to manage the contract with the consortium of SOMAFEL-BORIE/SAE-A.DEHE.

A factor which has had a great impact on the rehabilitation of the Railways in Mozambique has been the security situation. The impact has taken various forms such as

(*a*) delaying commencement of projects

(*b*) adding to work content by sabotage of track and bridges

(*c*) slowing down progress due to periods where works were suspended due to security risks

(*d*) general slowing down of project due to need for security escorts for work gangs. (These escorts were not always available in which case work did not proceed)

(*e*) additional direct costs to project of providing food for the protection force

(*f*) sabotage of locomotives which has made it difficult at times for CFM to provide sufficient motive power for works trains.

Rehabilitation by direct labour — Maputo to Chokwe main line length 194 km

Prior to rehabilitation of the track there were significant pre-construction activities to be undertaken. These included

(*a*) setting up the management team and equipping it with offices, transport and other logistics

(*b*) procurement of materials, equipment and plant

(*c*) rehabilitation of Umbeluzi Concrete Sleeper Factory

(*d*) construction of a new crushing plant at Estevel

(e) rehabilitation of Flash Butt Welding Plant at Machava

(*f*) setting up of a number of contracts for materials and services.

These activities took from August 1986 until May 1989, the last activity being the commissioning of the crushing plant in May 1989.

Umbeluzi Concrete Sleeper Factory was originally built in 1973 by Christiani Nielsen. It was taken over by the GOM after independence in 1975. It was producing 3000-4000 sleepers a month up to 1984 when the site was inundated by severe flooding and all equipment damaged and made inoperative. Extensive repairs were carried out in 1987 and new equipment imported. The cost of re-equipping the factory was £750 000 at 1987 prices. This excludes the technical supervision and labour. Later further equipment costing some £200 000 was provided. The procurement of the equipment, shipping, clearing from the port, erection of new equipment, repair of existing equipment and re-commissioning of the factory was completed in October 1988.

The sleeper production levels were low in the early days of production but increased to settle at approximately 8000 sleepers/month. The factory produced some 260 000 sleepers for the project up to February 1992 at which date requirements for the Limpopo Project were complete. An additional 90 000 sleepers for the Maputo-Chokwe section were imported.

CFM (Sul) did not possess a functioning quarry, producing ballast. The project provided a new set of crushing equipment and this was operated independently by BMS with technical assistance. This equipment was placed in an existing quarry with the quarry operators continuing to carry out the drilling and blasting. In hindsight the arrangement to rely on the quarry operators for broken rock was a mistake and very considerable delays to ballast production occurred. Only about 10% of the contracted rate of delivery of broken rock was achieved by the quarry operators. Later funds were provided by ODA to purchase drilling equipment and explosives to make BMS completely self-contained at the quarry. The rate of production of ballast and aggregates increased considerably thereafter. Quarry equipment and mobile plant was procured from the U.K. which cost approximately £1 695 000. Civil engineering works were undertaken by an International contractor at a cost of about £300 000. The procurement, manufacture, shipping, civil engineering works, erection and commissioning of the crushing plant was completed in May 1989.

In addition to the procurement from offshore there were a significant number of contracts to be negotiated and administered for locally produced materials and services. Generally these contracts were paid for in foreign exchange provided by donors. The contracts that were set up and administered included those

Executed in foreign exchange

- civil works at Estevel Quarry
- supply of fist sleepers
- supply of fist fastenings
- manufacture of workers accommodation huts
- manufacture of bridge deck units
- supply of cement from Cimentos de Mogambique
- supply of RHPC from Swaziland
- operation and maintenance of tamper/liners.

Executed in local currency (Meticais)

- manufacture of pandrol clips
- supply of rock at Estevel Quarry.

The track rehabilitation work on the main line covered 194 km of main line and 7 km of station and crossing loops. The work content included

(a) removal of existing sleepers
(b) re-sleepering with concrete sleepers
(c) re-ballasting — tamping and lining
(d) re-railing approximately 140 km of route
(e) thermit welding of LWR to CWR
(f) relaying all turnouts in main line.

Ancillary works included

(a) rehabilitating 30 km of sidings in Maputo Port
(b) relaying 63 turnouts in Maputo Port
(c) rehabilitating 7 km of branch line to the quarry
(d) rehabilitating 3 km of branch line to the sleeper factory.

The labour force utilised to carry out the track work was approximately 300 track men on the main line and 150 on the port works. These numbers do not include about 100 staff in Machava Depot, 140 at the sleeper factory and 40 at the quarry. Five permanent way supervisors were used, one in the depot, one on the port works and three on the main line works. The main line labour force was divided into two gangs who worked overlapping shifts of 20 days' continuous working followed by eight days' rest. The workforce camped out for their 20 day shifts at Magude (km 137) and earlier at Marracuene (km 35). The rate of sleeper changing varied greatly, from nil km in a month to 15 km in a month. Overall, from the date of starting to completion of sleeper changing, the average rate was 5.2 km/month. This is slow by normal standards but in the circumstances of a new organisation, untrained labour force, shortages of motive power and rolling stock, shortages of local funds, constraints imposed by the security situation and other problems it can be claimed that the work-force performed commendably in difficult conditions. For the period May 1991 to December 1991 the rate of sleeper changing averaged 10.2 km/month which represents more accurately what can be achieved if the logistics situation is normal.

The materials used for the rehabilitation included prestressed concrete monobloc sleepers with either Pandrol or Fist fastenings. Pandrol fastenings are the standard for CFM but Fist fastenings were used for the sleepers purchased from Zimbabwe as these were readily available. Where new rail was required BS 90 A rail section was used. This was welded into LWR at Machava and Thermit welded into CWR. Some 140 km of new rail was installed with the existing rail retained over the remaining length to Chokwe.

A significant amount of bridgeworks and repairs/extensions to culverts had to be undertaken. The work consisted of repairs to sabotaged structures where previous temporary repairs had been done, and re-decking of under-bridges to facilitate ballasting over bridges with previously open decks. Three bridges had to be reconstructed, eleven required new decks and two culverts required major repairs. Various other small repairs to bridges and culverts and the construction of ballast walls were carried out.

Consideration was given to undertaking the reconstruction of two sabotaged bridges by contract. Tender documents were issued to several contractors but only one bid was received. The bid valued the work at US$ 275 000 which was much higher than the engineers' estimate, even after allowing for the security risks involved. This high tender led to the decision to undertake the work by direct labour even though it was recognised that in house artisan skills were limited, as was the availability of construction equipment and materials. The reconstruction required the rebuilding of damaged piers, abutments and the provision of new wing walls. Simple designs and construction methods were adopted to suit the skills and materials available. Concrete blockwork was used for shuttering with mass concrete infill or concrete reinforced with rail where required. The blockwork was subsequently rendered to give an acceptable finish to the concrete work.

The re-decking of the bridges was carried out using pre-cast units. Two unit sizes were used with a length of 2100 m and 1400 m. The contract to manufacture the units was valued at £47 300 for 100 units of 2100 m, and 20 units of 1400 m. The concept of the units was based on similar units used previously by the National Railways of Zimbabwe. Structural design was undertaken by the consultants team in Maputo.

For straightforward re-decking a bridge closure of one day was normally necessary. For the reconstruction of the three sabotaged bridges the bridges were closed for several weeks and a deviation was built to allow the line to continue to operate. All the underbridges and culverts are provided for flood relief and are mainly concentrated between Magude (km 137) and Chokwe (km 213). Bridge repairs were programmed for the dry season of 1991, i.e. from April to November, the majority of the bridgeworks were completed in this period.

Quality of work achieved by direct labour

The quality of the track and bridge rehabilitation works met the specified standards. However, it is probably true to say that the completed work is not to the standard of finish that one would expect from an international contractor. This is particularly so for the bridge and structural works. It could be safely assumed that a contractor would have the skilled artisans and the necessary equipment to achieve a higher standard of finish. BMS were not in that position with either their skills or equipment. It can be argued, however, that a structure that is functionally safe but has some rough edges is all that is required in the bush.

Cost of rehabilitation carried out by direct labour

Table 1 sets out the costs of the rehabilitation and the unit costs for sleeper manufacture, ballast production, tracklaying, bridge repairs and the overall cost of rehabilitation of the track and bridges/km. Costs have been adjusted to 1991 rates. The capital costs of the equipment have been assumed to be written off over the life of the project.

Table 1. Track and bridge rehabilitation costs by direct labour, Limpopo Railway (adjusted to mid 1991 base)

	Equipment	Imported Materials	Local Mats. and Labour	Overheads Managemt & Super- vision	Total Cost	Quantities	Unit Cost
	£ '000	£ '000	£ '000	£ '000	£ '000		£
Sleeper Manufacture	1,200	2,400	481	1,200	5,281	260,000	20.3
Ballast Production	1,995	---	950	600	3,545	200,000 m^3	17.7
Tracklaying	1,937	1,230	1,550	4,000	8,717	201 km	43,368
Bridge Repair and Decking	300	---	52	200	552	N.A.	N.A.
Total Costs	5,432	3,630	3,033	6,000	18,095	201 km	90,025

Elapsed time to achieve rehabilitation by direct labour

Figure 1 shows in broad terms the programme elements from mobilisation in 1986 to completion in early 1993. It should be noted that

(*a*) procurement, including shipping clearance at arrival port and transport to site took a year or more for some major items, e.g. the Quarry Crushing Plant

(*b*) pre-track rehabilitation activities took 2.0-2.5 years from the instruction to mobilise

(*c*) sleeper changing took three years (average progress 5.2 km/month)

(*d*) the track rehabilitation, including bridgeworks, took 4 years

(*e*) overall the project required 6.5 years from mobilisation to final completion.

Training opportunities

Some formal and some informal training of the railway staff was undertaken. The formal training consisted of residential courses in the United Kingdom related to track and telecommunications for engineers and technicians, language training in an adjacent country and maintenance training in

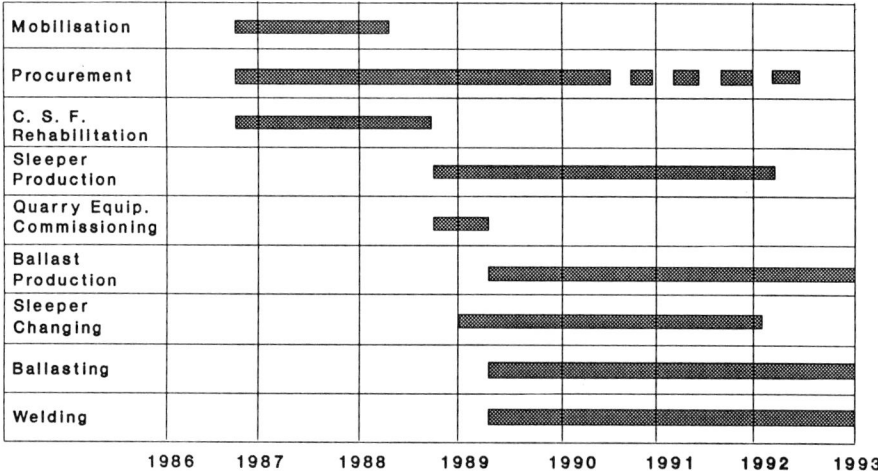

	1986	1987	1988	1989	1990	1991	1992	1993
Mobilisation								
Procurement								
C. S. F. Rehabilitation								
Sleeper Production								
Quarry Equip. Commissioning								
Ballast Production								
Sleeper Changing								
Ballasting								
Welding								

Fig. 1. Track rehabilitation by direct labour. Programme Maputo-Chokwe

the actual country. The informal or on-the-job training covered all aspects of the rehabilitation from project planning through logistics management to construction supervision and management. By the end of the project the clients' staff had gained considerable experience.

Financial contribution of railway organisation

The direct labour method requires a significant cost contribution by the railway to cover the direct cost of labour and their support costs. In the circumstances prevailing in Mozambique this can be an overwhelming burden and proved to be so on the Limpopo Project. The effect of a shortage of local funding was to slow down progress considerably.

Rehabilitation by contract with neighbouring railway organisation: Chokwe-Chicualacuala main line 258 km

Having negotiated a contract, mobilisation was the only pre-rehabilitation activity necessary. Sleepers were purchased from an existing commercial factory in Zimbabwe and a stock of some 30 000 sleepers already existed on site. Ballast was purchased from an existing commercial quarry in Zimbabwe.

The track rehabilitation work on the main line covered 258 km plus station and crossing loops. The work content included

(*a*) removal of existing sleepers
(*b*) re-sleepering with concrete sleepers
(*c*) re-ballasting, tamping and lining
(*d*) thermit welding of jointed track to CWR
(*e*) rehabilitating station yards.

The labour force utilised to carry out the work was approximately 270 track workers plus 2 supervisors. The shift work adopted was to work a 20 day continuous shift with the main force followed by 10 days rest. A small group covered the 10 day break to deal with emergencies. Temporary accommodation was set up at three locations along the line. The rate of sleeper changing averaged 9.2 km/month.

The materials used for the rehabilitation included prestressed monobloc sleepers with fist fastenings. New rail was not generally required but some spot re-railing was carried out to replace damaged sections.

There are no bridges on this length of the route and only a few small culverts. It was necessary to maintain an access track alongside the railway and to clear the bush from both sides of the track for a distance of 100 m for security reasons.

Quality of work by contract with neighbouring railway organisation

The quality of the completed rehabilitated track met the specified standards. Generally the standards achieved were marginally higher than those achieved by direct labour.

Table 2. Track rehabilitation costs by contract with neighbouring railway (adjusted to mid 1991 base)

	Works Cost £ '000	Overheads Management & Supervision £ '000	Total Cost £ '000	Quantities	Unit Cost £
Sleeper & Fastenings	6,797	1,184	7,981	422,400	18.9
Ballast Production	2,573	448	3,021	260,000 m^3	11.62
Tracklaying	9,873	1,720	11,593	258 km	44,934
Total Costs	19,243	3,352	22,595	258 km	87,577

Cost of rehabilitation carried out by contract with neighbouring Railway organisation

Table 2 sets out the costs of the rehabilitation and the unit costs of sleepers, ballast, tracklaying and the overall cost of rehabilitation/km. Costs have been adjusted to 1991 rates.

Elapsed time to achieve rehabilitation by contract with neighbouring railway

Fig. 2. shows in broad terms the programme elements from mobilisation in 1988 to completion in early 1992. It should be noted that

(*a*) there were no procurement or pre-track rehabilitation activities required

(*b*) sleeper changing took approximately two years (9.2 km/month)

(*c*) track rehabilitation took approximately 4 years to complete.

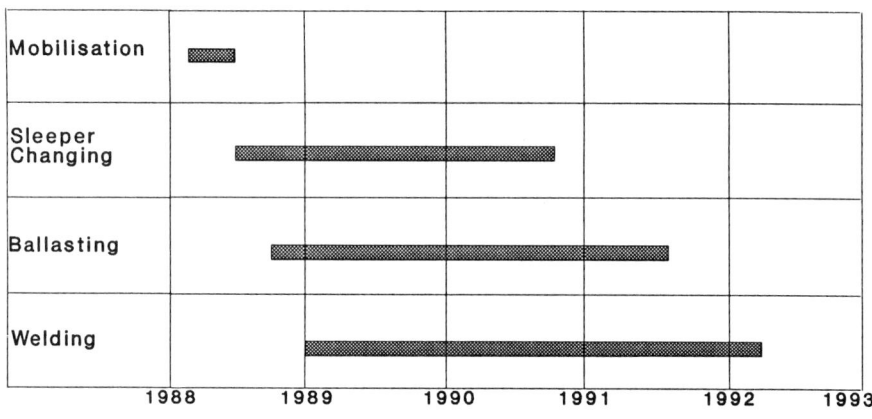

Fig. 2 Track rehabilitation by contract with neighbouring railway organisation: Chokwe- km 471

Training opportunities

A limited amount of informal training was included as part of the contract. This consisted of a group of 50 track workers being attached to the contractor's work-force and gaining on-the-job experience by being supervised by experienced railwaymen.

168

Financial contribution of railway organisation
The contractual arrangements did not require any contribution from BMS except the wages and support costs of the 50 workers attached to NRZ.

Rehabilitation by contractor. Nacalato Guamba main line length 533 km

For this contract a new sleeper manufacturing plant, a rock crushing plant and a flash butt welding plant had to be set up prior to tracklaying commencing.

The track rehabilitation work on the main line will cover 533 km plus station and crossing loops. The work content includes

(*a*) removal of existing sleepers
(*b*) re-sleepering with twin block concrete sleepers
(*c*) re-ballasting, tamping and lining
(*d*) re-railing with LWR
(*e*) thermit welding of LWR to CWR.

Data available on the labour force utilised is limited but it appears that 44 expatriates and 1250 Mozambicans are employed. The contractor elected to use mechanised tracklaying methods rather than the manual methods adopted on the Limpopo Line. He experienced problems with the maintenance of the equipment in a harsh working environment, and with the availability of fuels and consumables.

Table 3. Track rehabilitation costs by contractor: Nacala-Cuamba Railway (Adjusted to mid 1991 base)

	Works Cost £ '000	Overheads Management & Supervision £ '000	Total Cost £ '000	Quantities	Unit Cost £
Sleeper & Fastenings	27,102	3,335	30,437	900,000	34
Ballast Production	10,175	1,252	11,427	600,000 m³	19
Tracklaying	111,451	13,717	125,168	550 km	227,578
Total Costs	148,728	18,304	167,032	550 km	303,694

Cost of rehabilitation carried out by contractor: Nacala Cuamba line

Table 3 sets out the costs of the rehabilitation and the unit costs of sleepers, ballast, tracklaying and the overall cost of rehabilitation/km of line. The cost data is derived from the contract between the Consortium and the Government of Mozambique. The costs have been suitably adjusted to bring them up to a mid 1991 base and additional costs of new rail, technical assistance and other items outside the contract have been added to give an overall project cost. As the rehabilitation work is ongoing the costs quoted here cannot be regarded as being as accurate as those for the Limpopo Project. However, it is considered that the costs quoted are sufficiently accurate to allow a meaningful comparison with the other costs quoted in this paper. The costs given in the contract documents at 1/1/84 base levels have been increased by 40% to bring them to a mid 1991 level. The costs of providing and setting up the fixed plant and the cost of all mobile plant have been allocated completed to the first two phases. This plant will have some residual value at the end of the contract but this has been ignored.

Elapsed time to achieve rehabilitation by contractor

The contract programme provided 30 months for the first phase and 24 months for the second phase. The first phase was subdivided into 16 months for the preliminary works and 14 months for the rehabilitation between Nacala and Nampula (192 km). The first phase started in 1984 and was completed in January 1987, i.e. approximately in accordance with the programme. The second phase from Nampula to Cuamba was programmed to take 24 months but was interrupted by severe security problems and the contract was suspended for some considerable time. It has now recommenced with a programmed completion date of the end of 1993. During the first phase sleeper changing averaged 14 km/month.

Training opportunities

On-the-job training has been given to the staff of Mozambique Railways during the rehabilitation project. The areas of training have included surveying, track maintenance, track re-laying, finance and administration.

Comparison of rehabilitation methods

In a developing country the rehabilitation of a railway will almost always be dependent on external financing, either in the form of a grant or a soft loan. Thus there are two parties with an interest in how the project should be implemented, i.e. the railway authority and the financing agency. For the railway authority the objectives might be ranked as follows

(a) earliest availability of rehabilitated railway to maximise percentage of rehabilitation costs passing through national economy

(*b*) to minimise contribution from national economy

(*c*) to maximise training opportunities

(*d*) to minimise rehabilitation costs.

For the financing agency the objectives might be ranked differently as follows

(*a*) the most cost-effective method of achieving the rehabilitation

(*b*) to maximise training opportunities

(*c*) to minimise rehabilitation costs

(*d*) earliest availability of rehabilitated railway.

Clearly the priorities will often be incompatible and the decision makers will give a weighting to those aspects they consider the more important. With the financing agency this is likely to be the implementation method that shows the best cost/benefit ratios. This would often favour lowest cost and earliest availability of rehabilitated railway.

In Table 4 the unit costs of implementation by various methods are shown and in Table 5 the Author's own subjective views are given on the ranking of other aspects affected by the method of implementation.

Table 4. Unit costs of rehabilitation by various implementation methods (Costs given in Pounds Sterling at mid 1991 rates)

	Implementation Method		
	Direct Labour	Contract with Neighbour Railway	Contractor
Sleepers and Fastenings	£ 20 each sleeper set	£ 19 each sleeper set	£ 34 each sleeper set
Ballast Production	£ 18/m³	£ 12/m³	£ 19/m³
Tracklaying	£ 43,400/km	£ 45,000/km	£ 228,000/km
Overall Track Rehabilitation	£ 90,000/km	£ 88,000/km	£ 304,000/km

Table 5. Subjective ranking of the effect of the implementation method on aspects of the rehabilitation

Rehabilitation Aspect	Definition of Highest Ranking	Implementation Method		
		Direct Labour	Contract with Neighbour Rly	Contractor
Quality of completed work	Highest quality	3	2	1
First cost of rehabilitation	Lowest cost	1	2	3
Elapsed time to achieve rehabilitation	Shortest time	3	2	1
Training opportunities	Best opportunities	1	2	3
Financial contribution of railway organisation	Lowest contribution	3	2	1

When the cost information in Table 5 is being considered the following points should be noted.

(a) Sleeper and fastening costs for the direct labour method include all costs of the fixed and mobile equipment. This also applies to the contractor method but in the latter case the costs are spread over 900 000 sleepers whilst in the former they are spread over 260 000 only. It is probably reasonable to conclude that in the Mozambique context sleepers by the first two implementation methods cost sensibly the same with the sleeper cost by the contractor being 50% more.

(b) Ballast production costs for the direct labour method contain all costs for the fixed and mobile plant which are considerable (2 million). There is a significant useful working life remaining in this equipment. Also the production was impeded by actions outside the control of the direct labour organisation. For these reasons the unit cost is regarded as high and it is probably reasonable to conclude that in the Mozambique context the cost of ballast for the first two implementation methods is sensibly the same with the contractor's cost being 50% more.

(c) With tracklaying the costs by the first two methods are sensibly the same. The contractor's costs, however, are about 5 times the costs of the first two methods. Several factors could account for this large disparity

  (i) the contractor's inclusion of profit
  (ii) the mechanised rather than the manual approach to the work
  (iii) the enhancement of actual costs by a large factor to reflect the risks of undertaking a project through an underdeveloped country in an area which was not secure and where the works were subject to military attacks and to sabotage.

Conclusions drawn from comparison of the three methods

The following conclusions are those of the author alone and do not necessarily reflect the opinions of CFM, BMS or ODA.

(a) The costs and completion times for all three methods have been adversely affected by the need to rehabilitate the railways through areas subject to attacks by anti-government forces. For other developing countries with no security problems the costs should be significantly lower and completion times shorter.

(b) In the context of the situation in Mozambique the direct labour method is considered to be the best method if overall completion time is less important than lowest cost. If minimum completion time is significantly more important than first cost then the use of a contractor would achieve that result. The use of a contract with a railway from a neighbouring country has no significant advantages over undertaking the work by direct labour.

(c) In the context of a situation similar to that in Mozambique and where the railway serves an adjacent country it is concluded that the optimum solution is that adopted for the Limpopo Railway. That is to say a division of the project into two parts, one being undertaken by the neighbouring railway, which has a national vested interest in seeing the rehabilitation completed as quickly as possible, and the second part being undertaken simultaneously by the railway authority in the country concerned. This would retain the low cost advantage of direct labour and provide a reduced implementation time over a total direct labour method.

Acknowledgement

The Author wishes to acknowledge the source of the data used in this Paper as being reports and documents prepared for BMS, CFM and ODA . Any mistakes made in presenting this information are entirely the Author's.

Discussion on Papers 11 and 12

M. FULKER, Infrastructure Manager, Piccadilly Line London Underground Ltd
Would Mr Lindsay please expand on the following items that he briefly mentioned in his presentation Paper.

(*a*) How does he monitor performance against planning (e.g. method) and does he carry out self audits with relation to quality and standard of work?
(*b*) What criteria does he use (if any) when deciding to make or buy labour and plant.
(*c*) What methods does he adopt when planning cost-effective maintenance e.g. planned asset maintenance rather than expensive random asset maintenance

F. I. MAU, Vice-President Operations, BHP Rail Products (Canada) Ltd
Could Mr Masters say whether or not steel sleepers were considered for the Limpopo Line rehabilitation.

- The cost per sleeper would have been similar to the locally produced concrete sleepers, with the same fastening system
- The amount of ballast would have been less
- The cost of installation would have been substantially less — especially when transport and manual installation methods are considered.

P. E. COYSTEN, former London Underground, permanent way contractor
In the final analysis cost-effectiveness depends upon the operatives and their leadership. This is the same whether they be direct labour or contractors. There needs to be a clear set of objectives which are understood by all.

Staff selection and their training is therefore fundamental. Skills training, both theoretical and practical, based on task analysis has to be made available and I believe acknowledged by the achievement of a nationally recognized certificate.

Customer satisfaction is the basis of quality assurance and leadership training is designed to achieve this. Safety awareness and a relevant management system cannot be overlooked.

Competition in some form, either between gangs or districts or even within the competitive tendering process, plays a significant role in achieving the higher levels of efficiency.

Staff morale needs to be high and rising to achieve good performance and hence the facilities, tools, equipment etc. need to be provided. Equally, proper open lines of communication need to be maintained, which should allow for an overlap of the managment/staff structure so that senior management is seen to meet with operatives.

Staff need to feel that they have a role within their team's briefings to offer ideas for improvement in, for example, methods and equipment. It is not a bad idea for an operative to meet the supplier.

Full accountability for quality and costs by staff at all levels and rewards for the best or most improved, does help in the drive for a reduction in unit costs. Hence management information requirements will be sought by all concerned.

I have recently been involved with Hyland Joy and Wardrop and have become interested in their work on a number of computer based systems. Risk analysis in the determination of track standards, maintenance resource allocation models adaptable for different parameters; asset data base logging, modelling and analysis of historical records to assist in planning for quality and cost benefit are but a few which may be worthy of investigation by the railway organizations.

J. P. ORSI, Africa Regional Director, SofreRail

In Mr Masters' Paper about track rehabilitation in Mozambique, the figures given regarding the Nacala corridor project need to be corrected as follows

(*a*) unit cost for sleeper and fastening £30.00 instead of £34.00 (quantity 825,000)

(*b*) unit cost for ballast production £24.00 instead of £19.00 (quantity 950,000 m^3 instead of 600,000 m^3)

(*c*) unit cost for tracklaying £140,000 instead of £227,578.

The comparison with the Limpopo Corridor project has to take into account some important differences. The rehabilitation works from Maputo to Chokwe are located at a maximum of 194 km from the capital city and port of Maputo. The Nacala line is 2,000 km farther in the North, and extends 500 km through the bush. The logistic of this project is very heavy. We had to install camps with camp-bosses and doctors, import food and build houses for the Mozambican staff. We had to buy a plane which cost some US$ 2 million to the project. To face the very severe conditions of security, we had to employ eight security advisers, and to feed and equip 1800 Mozambican troops. The Mozambican railways (CFM) had also to armour their locomotives with the technical assistance of Transmark.

The Nacala corridor project includes an important part of civil works: rehabilitation of bridges and culverts, reinforcement of the formation, drainage and evacuation of zenithal waters, rectification of alignment with the objective of increased speed to 80/100 km/h.

However, this will not change the subjective ranking of the effect of the implementation method on aspects of the rehabilitation made in Mr Masters' statement, and I perfectly agree with it. But I would like to emphasize two aspects of the matter which have to be taken into account.

(a) The quality of the rehabilitation work is of paramount importance for a future cost-effective track maintenance.

(b) If rehabilitation is achieved in a shorter time, the railway will derive more revenue from the traffic.

This is important in an ecomonic analysis and will also be determining if the competition of other modes is in progress during track rehabilitation.

My opinion is that in most developing countries, the economic analysis will recommend the mechanized method. But it can work if there are locomotives to have the ballast trains, with drivers fed and paid and reasonable security conditions. In the Mozambican context, the difficulties encountered beyond the control of the operator did not allow full advantage to be taken of the mechanization.

A. J. J. LINDSAY, Author

My reply to the three questions by M. Fulker follow.

(a) Performance is monitored against plan at various levels within my organization.

I, and my senior management team, determine and monitor (both internal and external) key quality characteristics. These are chosen best to represent business requirements and the milestones to be achieved in meeting these goals. These indicators are compared with individual plans on a four weekly basis and used to highlight areas requiring management intervention. The indices are produced by my Quality and Safety Manager based on information supplied by the line managers responsible for delivery. Specific items are added or deleted from the list as appropriate throughout the year.

Examples of external quality measures are track quality indices, number of major workload items completed, unit costs of major work and aggregated train delays. Internal measures include safety hazards reported/rectified, rail defects found/rectified and on-track machine availability and reliability figures.

DISCUSSION

This strategic level of planning and staging is cascaded through the organization by the setting and monitoring of personal objectives to managers and supervisors. These specify the output required and the timescales and resource limitations to which work should be carried out.

Each major job, such as ballast or rail replacement is delegated to an individual who is responsible for delivery, cost, timescale and quality of finished work. The latter, however, can be difficult to specify in an easily measurable form.

Finally, normal day-to-day performance is reported via the CAMPS computer system. This details the percentage of the work plan completed and the amount of unplanned work undertaken together with the operator performances achieved.

Self checks are carried out at all levels and these not only examine the quality of work but also the method of working and compliance with the specified safety methods.

(b) My manpower numbers are planned in advance to correspond to my workload requirements and recruitment, if required, is staged to meet this plan. The forward plan is updated to reflect changes to working methods and workload. The use of contractors is governed by three main criteria: where the work is of a specialized nature and I do not have the in-house expertise; to cater for isolated peaks in workload primarily on structural maintenance and where I cannot compete economically with the external market rates.

The case for the purchase of new plant is based on normal investment criteria although this can be influenced by the cash flow situation within BR at any particular time.

(c) I am not too clear as to what is intended by this question. A proportion of my work is based on preplanned preventive maintenance cycles. Refettling switch and crossing layouts, rail joint attention and vegetation management are obvious examples of this. The majority of work, however, is determined by the various inspection techniques and prioritized according to engineering standards and business needs. In common with most railway organizations maintenance intervention levels are set to simplify and control this process. The general principle aimed for is 'just in time' maintenance.

In reply to P. E. Coysten, Mr Coysten has presented an excellent and concise statement on cost-effective maintenance.

I am somewhat cautious about the use of competition to generate higher levels of efficiency. The principle works if those involved feel that they can compete on equal terms. If this is not the case morale and performance can be adversely affected. I attempt to instill competition through comparisons with other civil engineers within BR and internally on specifically selected criteria.

A stronger motivator is, I feel, the desire to improve on curent performance and management must firstly be aware of such improvement and then
ensure that it is acknowledged.

I agree whole heartedly that staff selection and training are fundamental
to an organization's performance and that there is a prima facie case for
nationally recognized qualifications.

I was interested to hear of Hyland Joy and Wardrop's work. BR is also
developing similar models and undoubtedly this type of approach will
become more widespread in the future.

R. T. MASTERS, Author

In reply to F. I. Mau, both steel sleepers and timber sleepers were considered but rejected in favour of concrete sleepers. Given that a continuous
welded rail was to be installed I do not believe that steel sleepers would give
a technically satisfactory solution. Other factors in favour of concrete sleepers were

(*a*) the existence of a factory for the manufacture of concrete sleepers

(*b*) the opportunity to use locally produced sand aggregate and cement

(*c*) the opportunity to provide employment in the factory for Mozambicans

(*d*) the ongoing benefits to the Government of Mozambique of a factory
that could either continue to produce sleepers or be converted to produce
other pre-cast concrete products.

In reply to J. P. Orsi, I am grateful to him for the information he provides
on unit costs, and his endorsement of my ranking of the implementaiton
methods.

I would take issue with him, however, where he suggests there are
important differences in the logistical problems encountered in rehabilitating the Limpopo and Nacala lines.

Both projects deal with railways running through thinly populated bush
country. Camps had to be installed for both projects; food and houses had
to be provided for both projects. Security problems were encountered on
both projects. A similar number of Mozambican troops had to be fed on both
projects. However, locomotives were not armoured on the Limpopo project.
Here it may be of interest to note that the armouring of the initial locomotives
were carried out with the technical assistance of Mott MacDonald.

Transmark assumed the role of technical advisors in January 1991, prior
to which Mott MacDonald had been providing operating management and
technical advice to the Nacala railway from 1987 to the end of 1990.

13. Cost-effective maintenance of railway track — Indian Railways

M. RAVINDRA, Director, Indian Railways

Introduction

Indian Railways provide the principal mode of transport for freight and passengers, connecting people and places in the farthest corners of the country. They have been a great intergrating force since their inception in 1853. Indian Railways are a multigauge system with a total route kilometreage of 62,367 divided as follows

Broad Gauge (1676 mm)	34 880 km
Metre Gauge (1000 mm)	23 419 km
Narrow Gauge (762 mm and 610 mm)	4068 km
Total	62 367 km

There has been a tremendous upsurge in traffic on the Indian Railways, both freight and passenger — a trend which will continue in the foreseeable future. In the past four decades i.e. during the period 1950-51 to 1990-91, freight traffic has grown by 270% in terms of tonnage and 470% in terms of transport output. And as for passenger traffic, the growth has been equally remarkable — nearly 200% for the non surburban component and 450% for the surburban component. This increased traffic has been handled by gradually adopting such strategies as replacement of inefficient steam traction by diesel/electric, phasing out of conventional four wheeler freight rolling stock with bogie roller bearing stock, heavy haul operation etc. An idea of the growth of traffic density on the Indian Railways can be had from the graph given in Fig. 1.

Track on the Indian Railways has been traditionally maintained by manual means. The number of men required per kilometre is based on a semi-empirical basis, taking into account the traffic density, curvature, rainfall and condition of the formation.

The formula, originally evolved in the 1930s, has seen certain modifications over the years. Increased traffic density has, however, meant lesser availability of time for the men to maintain track. Coupled with this has been a reluctance on their part to do a very strenuous and monotonous task, a reflection of the changing socio-economic conditions in India. Absenteeism in gangs has been a major cause of worry for the organization.

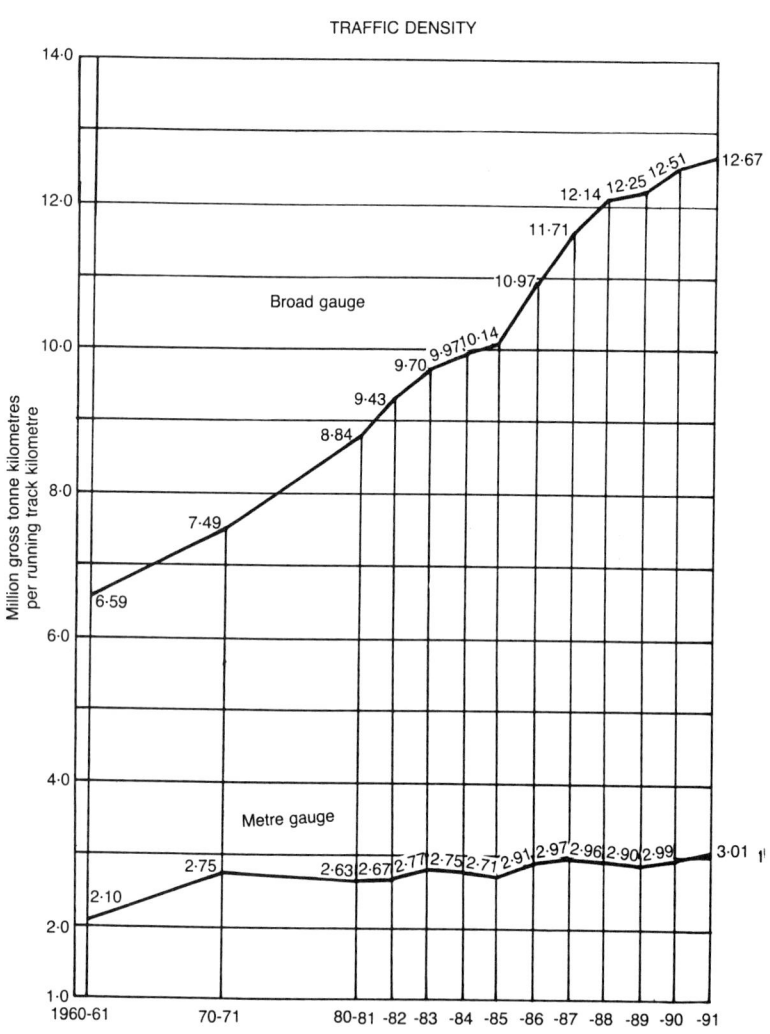

Fig. 1. Traffic density: transportation of one tonne of gross load including the weight of the rolling stock over one kilometre of running track

Strategy adopted for cost-effectiveness

To ensure a reliable track, safe for operations and at the same time cost-effective, Indian Railways have over a course of two decades adopted the following strategy

(a) laying down of better track standards

(b) gradual introduction of mechanized methods of track maintenance

(c) monitoring of track by track recording cars/oscillograph cars

(d) introduction of mechanised method of track renewals

(e) computerization of track management.

Track structure

For many decades, the track structure on trunk routes and main lines of broad gauge system of the Indian Railways comprised BS 90 R rail section, laid on wood, steel or cast iron sleepers to M+4 density; on the metre gauge, the corresponding standards were BS 60 R rail section. The depth of ballast prescribed was 20 cm.

With the main aim of providing a track structure having the least demand on traffic time for maintenance, a comprehensive review of the track standards for broad gauge was made in 1972. A similar exercise was done for metre gauge in 1979. As a result of the review, the broad and metre gauge systems on the Indian Railways were divided into 5 groups each, based on speed and traffic density. Apart from better maintainability, the track structure also ensured a longer renewal interval. The track standards were updated later, taking into account the introduction of heavier trains with higher axle loads.

On the broad gauge, for the heavy density routes with over 20 gross million tonnes of traffic, rails to UIC 60 kg/m profile and of 90 UTS are now the standard. For lower traffic density, 52 kg/m rails to Indian Railway profile, either 90 UTS or 72 UTS, are used. The sleeper density prescribed per kilometre varies from 1660 to 1310 per kilometre. Depth of ballast stipulated is 300 mm, the minimum being 250 mm.

On the metre gauge, the rail section now used is BS 90 R or BS 75 R, depending upon the speed and traffic density. The sleeper density prescribed varies from 1540 to 1230 per kilometre. The depth of ballast cushion varies from 25 to 15 cm.

Commencing from the mid sixties, there has been a gradual change over to the use of prestressed concrete sleepers with malleable cast iron inserts, elastic rail clips and pads. This change over has been necessitated due to economic considerations; a concrete sleeper is

(a) cheaper than cast iron, steel or wood with no environmental impact

(b) being heavier and with a longer life, provides a better and more reliable track at lower cost.

Indian Railways obtain at present over 4 million sleepers per annum.

The toe load of the elastic rail clips, originally 0.7 t to start with, has been increased to 1.0 t. The thickness of the elastic pads too has been increased from 4.5 mm to 6.0 mm. A clip suitable for use at joints has also been evolved. Concrete sleepers are now used at all locations — level crossings, bridge approaches and sharp curves. While the use of metal is being phased out, cast iron sleepers too are used in lesser number.

One feature which still eludes a satisfactory solution is using the same concrete sleeper for different rail sections. Liners of different thicknesses to accommodate varying rail foot widths is the solution adopted at present. The method, however, is not totally satisfactory as any missing liner or a wrongly used one can cause misalignment with consequent buckling.

Welding of track

As a measure of better maintainability and cost-effectiveness, Indian Railways commenced welding rails in the 1950s. Three methods are in use, these being

(a) aluminothermic welding
(b) flashbutt welding
(c) gas pressure welding.

Initially, the ordinary aluminothermic welding process with green moulds prepared at site was used extensively. In fact, a large majority of the field welds have been done by this process. Due to their relatively poor fatigue strength, weld failures are very common and are a matter of serious concern now. Since 1987, Indian Railways have switched over to the quick alumino-thermic welding process with short preheat, using air and petrol as fuel. The moulds used are of the prefabricated type. Further, the number of field welds are being minimized by close monitoring; for renewals flash butt welded panels, 130 to 260 m long are used. For rail or weld failures, wide gap welding is done.

Further improvement contemplated is the use of oxygen and liquid petroleum gas to reduce the preheating time.

As regards flash butt welding, seven of the nine zonal railways have captive flash butt welding plants. Two more are being added. Mobile flash butt welding plants are also available. Rails are welded in 10 and 20 rail lengths, i.e into 130 and 260 m long strings.

Indian Railways have only one stationary type gas pressure welding plant procured from Japan. The quality of welds produced by the plant has been excellent; there has been, till date, no weld failure. One reason for not adopting the process widely has been its low output. Trials to use the field method of gas pressure welding in new railway line construction is currently on hand as movement of welded panels is not possible at all locations.

Indian Railways commenced welding rails to five and ten rail lengths to eliminate joints in the 1950s; long welded rails were introduced as a regular measure in the 1960s. At present, the standards laid down are either long welded rails of maximum length 3 km with switch expansion joints at either end or short welded rails of 39 m length. The length of long welded rails has been restricted to 3 km and the use of switch expansion joints made compulsory after experiencing difficulties in destressing longer lengths. Track structure required, along with other parameters such as grade and curvature, have also been prescribed to ensure stability.

Improvements in turnouts

The turnouts used on the Indian Railways have been with over riding straight switches and built up crossings, laid on either wooden or steel sleepers. Such conventional turnouts have been a weak link in the track structure, needing frequent maintenance and replacement. Yet another unsatisfactory feature has been a severe speed restriction of 15 kmph permitted over the diverted route due to the large entry angle provided in the design. The first improvement effected was the use of curved switches with lower entry angles for turnouts located on the main lines, thereby minimizing excessive lateral forces. The use of 90 UTS rails for the manufacture of turn outs was also made obligatory.

Further, on high density routes, the use of cast manganese crossings is now prescribed with four times the life of an ordinary built up crossing.

Further improvements recently effected are

(a) the use of presetressed concrete sleepers for turnouts. The earlier design was the conventional one with sleepers at right angles to the straight track and in the crossing portion, to its centre line. As this meant having separate sets of sleepers for right-hand and left-hand layouts, a fan shaped layout has since been adopted. The same set of concrete sleepers can now be used both for left hand and right hand layouts, thereby leading to reduced manufacturing costs and inventory

(b) a design which permits the use of BS 90 R, 52 kg/m and UIC 60 kg/m rail section. There is no need to change the sleepers. The different foot widths of the rail sections are accommodated by liners of different thicknesses.

(c) use of elastic rail clips

(d) use of rails with thicker webs for switches to obtain greater lateral strength.

Track maintenance

As mentioned earlier, the track on the Indian Railways has been traditionally maintained by manual methods. Out of a total regular employee strength of 1.65 million, those engaged on track maintenance alone consitute 20% of the employee strength, that is nearly 0.3 million staff. The wage bill of the Indian Railways constitutes nearly 50% of the total expenses and it was, therefore necessary to make track maintenance cost-effective. Changing socio-economic conditions in the country and reluctance on the part of men to work on track, heavier traffic density and use of concrete sleepers in larger measure were other compulsions which required the introductions of mechanization for track maintenance.

Heavy duty on track tie tampers were introduced on the system in the early sixties. The number has increased progressively. And the type of machines procured has also undergone improvement. The latest series of machines procured are of the continuous action type which give better productivity and higher quality.

Other type of track machines introduced are tie tampers for turnouts, dynamic track stabilizers, ballast regulators and ballast cleaning machines. As regards ballast cleaning, the periodicity adopted on the Indian Railways is about 15 years. The method most widely used is manual. A track with concrete sleepers does not any longer permit manual means to be used. Hence the switch over to ballast cleaning machines.

One of the major constraints faced by the Organization in track maintenance is non availability of adequate traffic blocks, consistently. The pattern of traffic obtaining and intensive utilization of track are the prime reasons for this.

Changes in maintenance organization

Indian Railways will soon be switching over to a three tier system of maintenance on heavy density broad gauge routes and to a two tier system on broad gauge routes with lesser density and on metre gauge. In the three tier system, the first tier would comprise through packing of the track by heavy duty on track tampers at prescribed intervals. A mobile mechanized unit, equipped with off track tampers and other hand tools including welding equipment would constitute the second tier and deal with stretches of track needing urgent attention. The mobile mechanized unit would also deal with rail and weld fractures. Stationary gangs of reduced strength would constitute the third tier and deal with such routine work as patrolling, greasing of fish plates and elastic rail clips etc. The system is at present on trial on the zonal railways and is expected to rationalize track maintenance operations considerably when adopted as a regular measure.

In the two tier system, mobile mechanized unit would do regular through packing too; heavy duty on track tie tampers would not be used.

Progressive mechanization of all track maintenance operations by having a judicious mix of high power automatic on track tamping machines, light on track tampers, off track tampers, small hand operated power tools etc., is the aim.

Ultrasonic railflaw detection

To avoid in service rail/weld fractures, ultrasonic testing of rails by manually operated equipment is done to schedules laid down. The schedule is based on the section of the rail, cumulative traffic carried and the traffic density. All rails before being laid in the track are extensively tested. All welds are also similarly tested. Indian Railways have one Self Propelled Rail Testing Car (i.e. SPURT) which is used on routes with concrete sleepers and long welded rails, i.e. where rails do not have fish bolt holes.

Re-laying of track

Another area where cost-effectiveness together with quality has been brought about is in the re-laying of track. The traditional method of re-laying has been by manual means. With the introduction of concrete sleepers and heavier rails, an alternative mechanical method had to be developed.

Panels with service rails are preassembled in a depot and laid by the portals; welded rails, used for linking the auxiliary track, then replace the service rails. The system has been in use for over two decades on the Indian Railways. Re-laying is now mostly done using portals working on an auxiliary track.

Indian Railways have also acquired two track re-laying trains. These trains are capable of lifting the old track and placing the new track in position, all automatically. No released material is left at site, thereby eliminating the need for subsequent traffic blocks for picking up the released materials. Productivity is relatively high, as also the quality of track linked.

Track monitoring

If track is to be maintained well to ensure safety and reliability of operation, monitoring at regular intervals is a must. Indian Railways were earlier dependent on the Hallade Apparatus — an instrument that was used for many decades.

Track Recording Cars using mechanical means for measurement were introduced in the early 1960s. These were later changed to electrical and electronic systems respectively. A further development has been the use of inertial principle for measurement of track parameters.

The standard deviation concept to measure track irregularity has also been introduced; a length of 200m has been adopted for calculaiton purposes. The field staff get indication as to when track requires to be attended to immediately or later. On the trunk routes, the cars are run once every three months.

Oscillograph cars and portable accelerometer cars are also used at periodic intervals (once every 4 months and once every month respectively) to judge the quality of track.

Development of track management system

Quite a few of the railway systems have already developed the use of computers for track maintenance. MARPAS developed by BR is one such example. A computerized Track Management System not only indicates the present state of the track but also forecasts likely degradation to enable optimal deployment of tie tampers. It also helps the management decide as to whether a relaying must be done or more intensive maintenance. Interruption to traffic due to working of tie tampers can be reduced by attending to track only when needed and not on a calendar system of maintenance. The number of tie tampers required for a system can be optimized, as also the use of other resources.

Indian Railways have just commenced the development of a track management system. A trial on 100 kilometres of track on all the nine zonal railways has been planned; the system is already operative on Northern Railway. It is expected that once the system is in operation on a regular measure, substantial savings would be possible in many areas such as

(a) postponed renewals
(b) better utilization of tie tampers
(c) lesser intervention to traffic etc.

Other measures in hand

Other areas where Indian Railways are actively engaged for economizing on track maintenance and renewal costs are

(a) grinding of rails for removal of surface defects and for a better rail-wheel contact. One grinding machine is on use on the Indian Railways.

(*b*) revision of the wheel profile; the profile mostly used is a straight tapering cone, the taper being 1 in 20; a profile based on wear for freight stock is on trial now. It is seen that adoption of the "worn profile" gives a much better rail-wheel contact.

(*c*) revision of periodicity prescribed for deep screening of track. A more rational method for complete screening of ballast is being worked out.

Epilogue

Indian Railways have been continuously upgrading their technology for the laying, monitoring and maintenance of track with a view to making it not only more reliable but also cost-effective. The technical changes brought about have, in turn, meant training of the work-force at all levels; assimilation and acceptance of the change are otherwise not possible. The training institutes of the Indian Railways and the Research Organization of Lucknow have played an important role in this regard and continue to do so.

14. Cost-effective maintenance of railway track — Scotrail

R. M. CHORLEY, BSc, MICE, Civil Engineer, North Scotrail

Introduction

This Paper deals with the maintenance of permanent way in the North area of Scotrail. All areas of British Rail can claim to be unique as all enjoy (or suffer from) particular features, geography or conditions not experienced elsewhere. I begin with a broad picture of the three distinct types of railway in the area before focusing into detail on individual elements that make up the requirements and cost of maintenance.

(*a*) The area includes the northernmost 122 miles of the East Coast Main Line Al and 45 miles of the West Coast Main Line both being 100 miles/h railway and with a mix of Intercity, Regional Rail and Freight Sector trains.
(*b*) The railways from Perth to Inverness and Aberdeen to Inverness are generally 75 miles/h single lines with passing loops mostly Regional Railways but with some InterCity and Freight Trains.
(*c*) Beyond Inverness the lines to Kyle of Lochalsh in the West and Wick and Thurso in the North total 238 miles of lightly used single line Regional Railways with seasonal InterCity Tourist Trains and occasional Freight.

Therefore to maintain the infrastructure to the standards referred to in a previous paper at this Conference the area has an annual budget shown overleaf.

Permanent way maintenance 1991/1992

Money spent	£5 384 000
Route miles	714
Single track miles	957
Units of P & C	499

i.e. 5600 per single track mile (including $^1/_2$ lead)

Let us see how the 5600 per mile is spent and then consider each component part and evaluate more economical methods. I have split this into ten headings by percentage in Table 1.

Table 1. Budget breakdown by activity

Work	Item %	Cumulative %
Fencing	13	13
Banks, bushes, trees, weed-killing	11	24
Access gates, steps, cesses, drainage	8	32
Track patrolling, rail flaw detection	10	42
Sleepers, spot resleeper, regauge	8	8
Fastenings, keys (plain line and P & C)	8	16
Fishplates, expansions, destressing	10	26
Rails: turn, transpose, change, reprofile	5	5
Track geometry maintenance-manual	19	24
Track geometry maintenance-mechanical	8	32

I shall analyse the build up of these ten broad categories of work and suggest present and possible future methods of achieving the required standards at an economic cost. But first I must comment on the relative proportionality of these expenditure headings.

Off-track work now takes about one-third of the allocated budget and this is dictated not by the business requirement of the Railway but by the law of the land. Examples include

- The Railway Clauses Consolidation Scotland Act 1845 – liability for lineside fencing, drainage
- The Railway Clauses Consolidation Act 1845 – dimensions of private/public structures
- The Roads and Bridges Act 1878 – responsible for maintenance
- The Trunk Roads Acts 1936 and 1945 – ownership
- The Roads (Scotland) Act 1884 – transfer of ownership to Secretary of State
- The Railway Clauses Consolidation Scotland Act 1845 – responsibility for maintenance and dimensions
- The Level Crossing Act 1983

A further 10% expenditure, is in BRB requirements for patrolling, rail flaw detection etc. This is an essential safety requirement but note that we have now spent 42% of our budget before we can permit the first train to run on the railway track.

The next three headings of expenditure are partially but not wholly caused by the running of trains and total 26% of our budget.

You will note that we now have less than 30% of our authorised budget to spend upon the geometric rectification of alignment and top where these have deteriorated below the required standard for the safe running of trains.

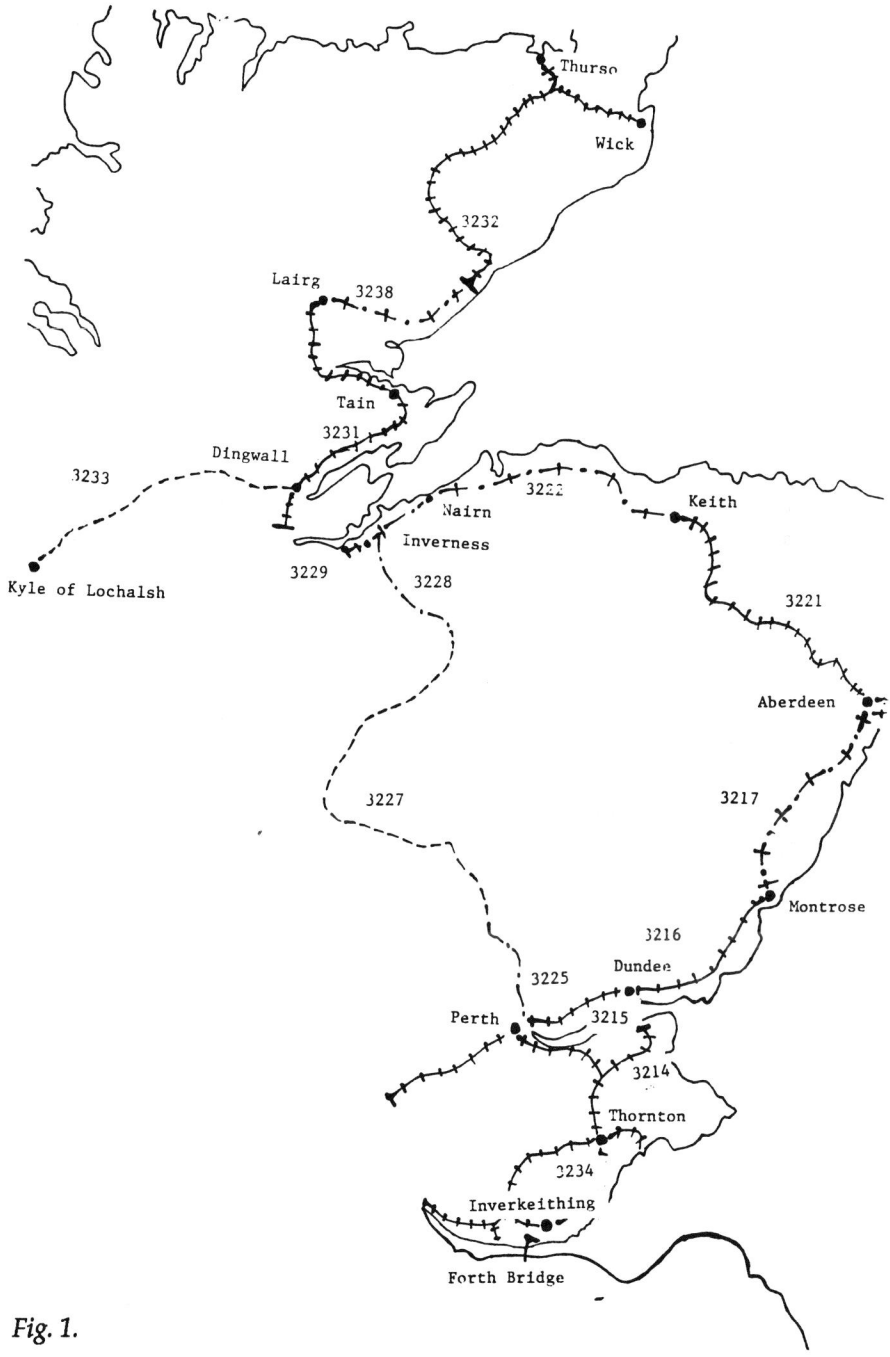

Fig. 1.

Therefore any approach to economic maintenance that concentrates solely on rectification of track geometry is tackling less than one third of maintenance expenditure. Now let us examine in detail the ten headings.

Fences etc.

I can see no circumstance in which a UK Parliament would relax the requirement for British Rail to fence our railway routes. A new or replacement fence costs an average £11 000 per mile of fence or £22 000 per mile of route - no matter if the railway is single, double or quadruple track. At our current rate of expenditure fencing has to last on average 30 years between complete renewal. I consider that there are 3 types of fencing requirements

(a) *Child fences*. In built-up areas we need a better standard of fencing both to keep children off the railway for their own safety and to keep vandals from inflicting damage to the railway and trains.

(b) *Cow country*. George Stephenson was asked at a House of Commons Select Committee what would happen if one of his steam locomotives hit a cow. He replied 'So much the worse for the cow' and for about 100 years he was right. But in the UK, (and colleagues in Ireland and India have told me of similar problems) trains are now being derailed by cattle on the line. I believe that this is for two reasons — cattle breeding has produced bigger, heavier cows and mechanical engineers have produced lighter, multiple unit rolling stock. I believe our standard of this type of fencing is adequate but more attention must be given to inspection and speedy repair to defects.

(c) *Sheep country*. The standard of fencing to make a rural railway completely sheep and lamb proof would be prohibitively expensive. In my experience, sheep always get on the line and the fencing keeps the animal on the line until it is knocked down by a train. I believe the Railway should negotiate with the larger landowning neighbours an agreement that the landowner receive an annual payment and accept responsibility for maintaining the fencing to the standard he requires and he stands the cost of animals lost.

Banks, bushes, trees and weed-killing

Trees are emotive and the civil engineer is in trouble whether he does nothing or does something. We need positive landscape management of land within railway fencing by planning suitable treatment for every route mile in consultation with environmental interests followed by professional treatment with appropriate practices and modern herbicides to achieve and maintain. This will require capital expenditure before resultant savings can be made in its category of work.

Access gates, steps, safe cesses, walkways and drainage
This is an ever increasing workload to achieve the standards required by the Minister of Transport Safety Inspectorate. I am not complaining. I believe we need higher standards of safety for our railway staff and return to this theme in my conclusion.

Track patrolling
Introduction
Traditionally permanent way has been patrolled/examined/inspected by a man walking the track and

(a) noting and reporting defects
(b) carrying out minor repairs
(c) if necessary, stopping or cautioning trains.

The present position
All lines on this area are patrolled by foot at frequencies stipulated by the BRB manning formula (714 route miles).

In January 1988 all Patrolmen's record books covering the 238 route miles North and West of Inverness were called into my office. They were examined and the type and number of reported defects were analysed and totalled into groups.

The number of Patrolmen at present used on the 714 route miles is 32 Patrolmen and 32 Lookoutmen.

The patrolling machine
A Permaquip inspection trolley was delivered to the CEN Perth in December 1987. Some problems were encountered and overcome in commissioning the machine and training staff over the Christmas-Hogmany period but by mid-January trial running commenced.

Further problems were experienced when the Buggy occasionally failed to activate track circuits at Automatic Barrier Crossings. The Plant Section (CE) were able to design and fit a shunt device that proved satisfactory in service.

The first trial period
During the three months January to March (when difficulties were being overcome) a few test runs were made to assess what could and what could

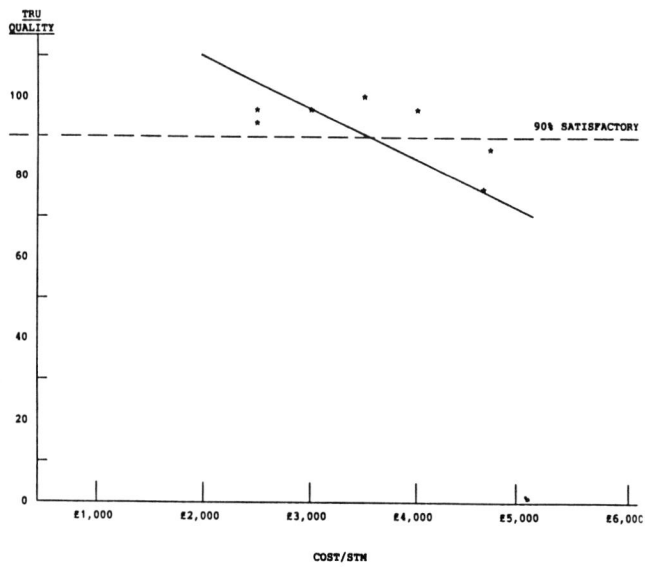

Fig. 2. Quality v. maintenance costs per STM

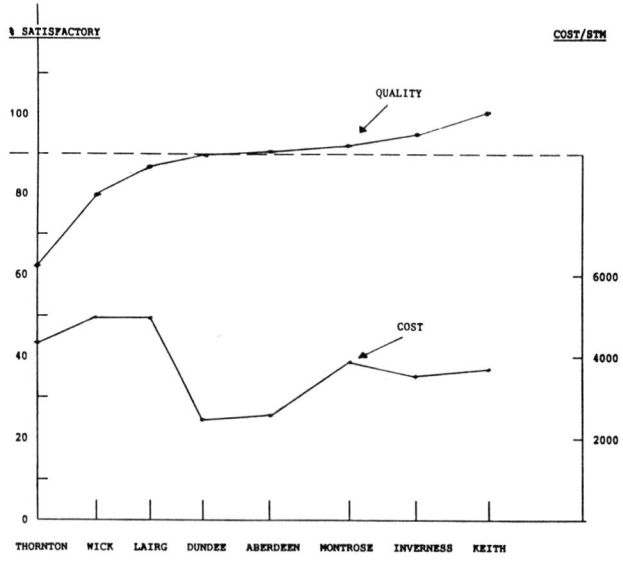

Fig. 3. Quality v. maintenance cost by business route sectors

not be seen by an observer travelling on the machine at a range of different speeds.

In April and May the machine was taken throughout the PWS sections Dingwall, Tain, Wick and Inverness (part) to familiarise permanent way staff with the machine and explain what was expected of it.

In June the machine was allocated to Wick PWSM at Georgemas Junction to carry out comprehensive comparative testing of the two methods of patrolling, i.e. by foot or by machine.

Now all normal patrolling on this Ssction is being carried out by two leading trackmen in the machine supplemented by independent foot inspections by Track Chargeman and Supervisor.

Patrolling speed

10 miles/h. The first tests were at l0 miles/h and it was surprising just how much detail of individual fastenings, keys, and defects can be seen at this

Table 2. Summary of results

Mode of Patrol / Defect	Foot 2mph	Machine 10mph	Machine 20mph	Machine 30mph	Machine 40mph
Long Top	1	1	1	1	1
Short Top (joints)	1	1	1	2	2
Alignment	1	1	1	1	1
Fastenings	1	1	1	0	0
Keys	1	1	1	0	0
Fishplates Broken	1	2	2	2	2
" Cracked	1	0	0	0	0
" Loose	1	2	2	2	2
Voids	0	2	2	2	2
Ride Quality	0	0	2	2	2
Ride Defects	1	0	0	0	0
Cesses	1	1	1	1	1
Banks	1	1	1	1	1
Fences	1	1	1	1	1

0 means cannot be detected by this mode of patrol
1 means can be detected by "eye" by mode of patrol
2 means can be detected by "feel" by mode of patrol

197

speed. This is because the observer is sitting at a low comfortable convenient position able to devote 100% attention to examination.

20 miles/h. At first more difficult but once the eye has become accustomed to the faster pace (like a batsman in the first few overs against a fast bowler) most features can still be seen at 20 miles/h.

30 miles/h. Individual detail and fastenings become a blur though observation of line, top, quality of ride, etc. still effective.

40 miles/h. Long top, alignment, ballast conditions, cesses, banks and fences can still be seen satisfactorily at this speed. Also gives good indication of riding quality of track.

50 miles/h. At particular locations, e.g. P & C, longitudinal timbered bridges, level crossings, and particular trouble spots the machine is stopped and both driver and patrolman examine and carry out work if required . NOTE : They do such work safely under the complete track possession required by the presence of the machine on the track.

Summary of results

I have pondered long how to present the results in some numeric form based on a scale for each feature of 1 to 5 but the results on paper look confusing so I have devised a simpler matrix showing the list of defects we look for against the band of patrolling speeds (see Table 2).

From this table it can be seen that the only features that cannot be detected from the machine are cracked fishplates and rail defects (NB. There were 16 reported during 12 months in 1987 on 238 single track miles).

On the other hand the machine method can detect voids and ride quality that the foot patrols miss.

Although quality has not been included in this matrix it has been quite obvious to any observer or participant in these trials that a patrolman sitting in a comfortable position immediately above the rail is able to devote 100% attention to the examination in hand. By comparison, even the most vigilant patrolman by foot has the primary consideration of where to put his feet, and watch out for trains, also weather conditions and other distractions. Quality of examination is and should be far greater by machine than by foot.

Safety for a patrolman by machine is, of course, considerably greater than by foot.

Remedying defects

Immediate action. Stop and repair. There are always two men (driver plus patrolman) and the facility to carry on the trolley more tools and materials than can possibly be carried by the most superhuman patrolman. Also considerably safer as trolley ensures possession of the line.

Intermediate action. Report to PWS Office (by radio or telephone) and judgement used whether to repair on return journey or send gang.

Longer term action. Included in CAMPS input.

Applying TSR or other emergency. Obviously this can be done far more effectively from a trolley in section with modern radio or telephone communications than with a man with flags and detonators.

Sleepers

On lighter used railways we no longer carry out complete renewal and replacement with continuous welded rail and concrete sleepers but by planned patch resleepering. This is not carried out by resleepering every third sleeper but by careful examination and only changing rotten sleepers as indicated by sleeper integrity testing machine.

This may be as many as ten in one 60 ft length of track or as few as two in another length but usually averages 30% over a quarter mile section. As the average life of a sleeper on these routes has been 45 years, this provides reconditioned track capable of another 15 years life without major expenditure until the next cycle of patch resleepering.

Fastenings

There are many types of fastenings available in the world. The most cost-effective are those that feature "fit and forget" characteristics and not conventional bolts that work loose.

Rail joints

Fishplated joints are costly to maintain and become a potential point of breakage at the rail bolt holes. It is also extremely difficult to oil (to allow for expansion) even 50% of the plates between the last of the winter snow (often April) and the first high rail temperatures (often May). We are now starting to cut off the ends of rails and welding in situ to provide serviceable bullhead welded track.

Relative costs (all per single track mile) are

New concrete sleeper	£260k
Serviceable concrete sleepers	£220k
Controlled patching 30%	£40k
Plus rerailing (if required on curves)	£16k
Plus cropping rail ends and welding	£75k

Rails

A bad engineer changes rails after they break in the track. A poor engineer changes rails long before they would have broken in the track. However, a good engineer changes rails just before they would have broken in the track to obtain the most economic, useful life out of a rail.

Rails break because of fatigue (which is load plus frequency). There are 3 trains per day beyond Inverness with 18 ton maximum axle weight. Therefore, by special dispensation from the DoCE we are allowed to have an increased wear of 2 mm of overall depth of rail before relaying is required. This means 135 mm depth reduced to 133 mm on the far North line and 132 mm depth reduced to 130 mm on the Kyle line (D Category). In the unpolluted atmosphere of Northern Scotland and light traffic, the 2 mm provides 20 years additional wear of the rail. This reduction also allows us to replace sideworn rails on curves by bullhead rail cascaded from relaying on the Perth to Inverness and Aberdeen to Inverness lines. The accountant charges us scrap price and as we produce scrap rail out of the existing track, the cost of rerailing is at the cost of labour and engine power only.

Cheap and cheerful CWR

This is a subjective type of judgement but with differing types of rail rolling stock being introduced throughout the world from heavy axle and high frequency lines to light transit systems and infrequently used rural railways, may I make a plea for permanent way engineers of all types to co-operate together and use rails economically and extract full value from them. Thus I can foresee economic benefits to all types of railway if

(a) a typical rail spent 15 years in a 125 miles/h or heavy axle route
(b) after refurbishment by grinding or planing the running surface it would serve a further 25 years in a lighter used 90 miles/h railway.

After further reprofiling of the running surface, ultrasonic testing and repair, by welding, of any defects the rail would be put into a 50 miles/h branch line for a further 25 years. This is what we are now beginning to achieve but why not another step?

At this stage the rail would still retain sufficient weight and strength for rerailing some of the light rapid transit railways now being built or on the drawing board and last in these conditions for a further 35 years.

If we could collectively as permanent way engineers, (and not be fragmented by our political and business masters), consider a potential rail life of 100 years and financial appraisals taken into account on discounted cash flow the re-sale and re-use on at least four occasions, the economic equations of both renewal and maintenance costs of existing railways and the capital cost of new railways could be transformed. It would require a matching commitment by railway plant manufacturers and component suppliers to provide machines and components to allow the economical refurbishment and re-use of rails and also fixing components.

Manual geometry maintenance

Even small errors in alignment or top induce increased forces on the track which consequentially causes further deterioration. Effective and economical maintenance is based on

(a) early detection of track faults

(b) quick and effective remedial action by mobile gang suitably trained and equipped.

To achieve quality and safety an increasing proportion of both inspection and manual repairs is being carried out under track possession — about 65% in the first quarter of 1992 on the North Area.

Machine geometry maintenance

The cheapest method of maintaining track is to install to a high standard of geometry and react quickly to defects as they occur and rectify them before they develop into larger sections of poor track. Lowering standards is false economy. High quality track is cheaper to maintain than bad track as well as being better for both passenger and rolling stock. My area is split into 29 Business Route Sectors and my maintenance costs are low where track recording unit shows good track and costs are high where track quality is poor.

Conclusion

Permanent way is not eternal but has a life of about 30 years. We should now be preparing our track and ourselves for the 21st century. Picking up the threads out of the 10 elements of maintenance, I conclude that

(a) for each route it should be decided what type and standard track, banks, fencing are required

(b) maintenance requirements should be assessed, i.e. labour, materials, machinery, training and working methods

(c) every job, however minor, needs planning and appropriate resourcing

(d) for safety and quality we must actively plan that the majority, and eventually, all work on railway track is carried out under planned controlled possession of line conditions.

15. Cost-effective track maintenance on Queensland Railways

F. BELL, Dip.Eng, MIE Aust.,CPEng, FPWI

An overview of Queensland Rail

Queensland Rail (QR) is a narrow gauge (1067 mm) system of some 10 000 km. Although it is a Government owned system, during the past two years it has undergone massive restructuring and conversion from a Government Department to a commercially-driven Government-owned Enterprise. This change will result in QRs corporatisation during this year and to all intents and purposes, a total withdrawal of Government funding.

QR began its life 125 years ago as a series of short developmental railway systems providing a service from the many ports along the Queensland Coast. It was not until 1923 that the systems were joined to form a system very much as exists today. These early lines were of small rails (40 60 1b/yd) dog-spiked directly to timber sleepers. The formation followed the natural contours to a large extent and the ballast was sand or gravel. Bridges were generally of timber construction except at major river crossings where steel and concrete were used.

Many of these original lines are still in service, still with the original rails and still with gravel ballast. The more important lines have undergone upgrading or reconstruction, however the predominant track structure is still quite light with 41, 47 or 50 kg rail on timber sleepers. More recent construction and reconstruction incorporate concrete or steel sleepers and rails of 50 or 60 kg/m.

In the late 1960s, Queensland's major coal deposits were developed and expanded to meet an increasing world demand for both coking and steaming coal. During the following 20 years some 1000 km of new track was constructed to a heavy haul standard to undertake the transport of the coal to three major coal port developments. Trains of up to 11 000 tonnes with 5 or 6 diesel or electric locomotives and a length of over 2 km are used in this traffic. Wagons of up to 100 tonnes are used. At this time QR moves in excess of 1 000 000 tonnes of coal each week. The track is predominantly 53 or 60 kg/m rail on concrete sleepers.

It is this coal traffic along with the haulage of other minerals and mineral products that has provided QR with the opportunity to become a fully commercial organisation.

The cost of track maintenance

The range of track construction standards in operation within QR has provided the opportunity to look at maintenance costs as they vary with both the track standard and the traffic task. QR has three basic track standards related to the axle loading of the traffic operating over them. They are as follows

B class	less than 15.25 TAL
A class	15.25 to 18.5 TAL
S class	greater than 18.5 TAL

This standard is in general controlled by bridge strengths. The engineering standards for the track are more comprehensive and are shown in Fig. 1.

In monitoring the track maintenance costs QR records the cost for each individual line section. When this cost, expressed as $ per kilometre, is related to gross tonnes over the line section, a graph as shown in Fig. 2 can be plotted.

In this case the X axis represents the length in kilometres, the positive Y axis represents gross tonnes and the negative Y axis track maintenance cost in $ per km.

This information is quite useful in relating maintenance cost to traffic task. It clearly shows how the higher standard of track in the heavy haul line has a lower maintenance cost even though the traffic task is high.

TRACK STANDARD (1)	RAIL SIZE (kg/m)	LENGTH (m) (2)	SLEEPER TYPE (3)	SPACING (mm)	PLATING (4)	BALLAST DEPTH (mm)	COMMENT
20.1	20	short	T_2	685		100	
31.1	31	›110	T 1/2	685	R<240 m (outer rail)	150	
31.2	31	›110	T_1	685	R<400 m	150	
41.1	41	›110	T_1	685	R<400 m	200	
41.2	41	›110	T_1	685	fully plated	200	
41.6	41	CWR	C	685	–	250	
50.1	50	›110	T_1	685	R<400 m	200	
50.2	50	›110	T_1	685	all curves	200	
50.3	50	›110	T_1	685	fully plated	200	
50.4	50	CWR	T_1	610	fully plated	200	
50.5	50	CWR	T_1	610	fully plated	200	
50.6	50	CWR	C	685	–	250	
50.7	40	CWR	S	685	–	200	
50.8	50	CWR	C	685	–	250	Dual Gauge
53.1	53	CWR	T_3	610	fully plated	250	
53.2	53	CWR	T_3	610	fully plated Pandrol on curves	250	
53.3	53	CWR	C	685	–	250	
53.8	53	CWR	C	685	–	250	Dual Gauge
60.1	60	CWR	T_3	610	Pandrol	250	
60.2	60	CWR	C	685	–	250	

Fig. 1. Engineering track standards

Fig.2. Track maintenance costs v. tonnage

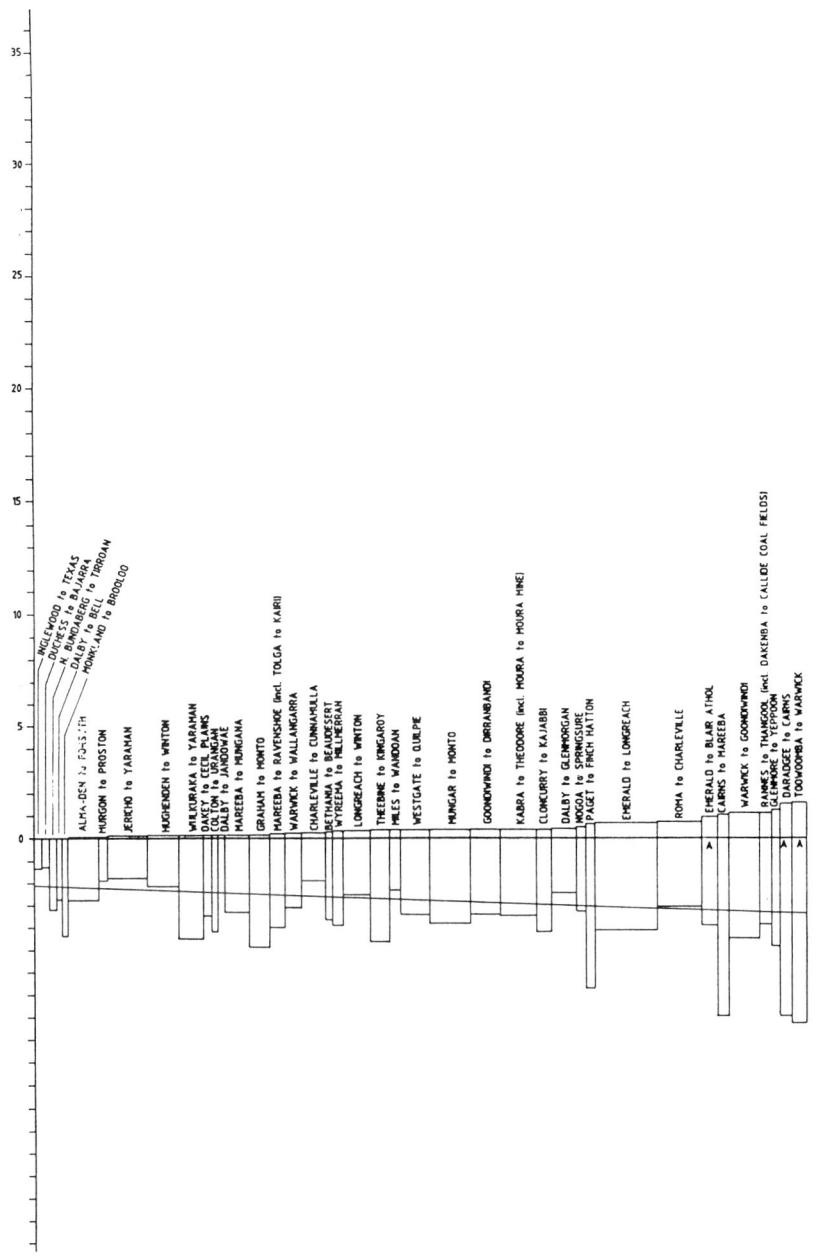

Fig. 2. contd.

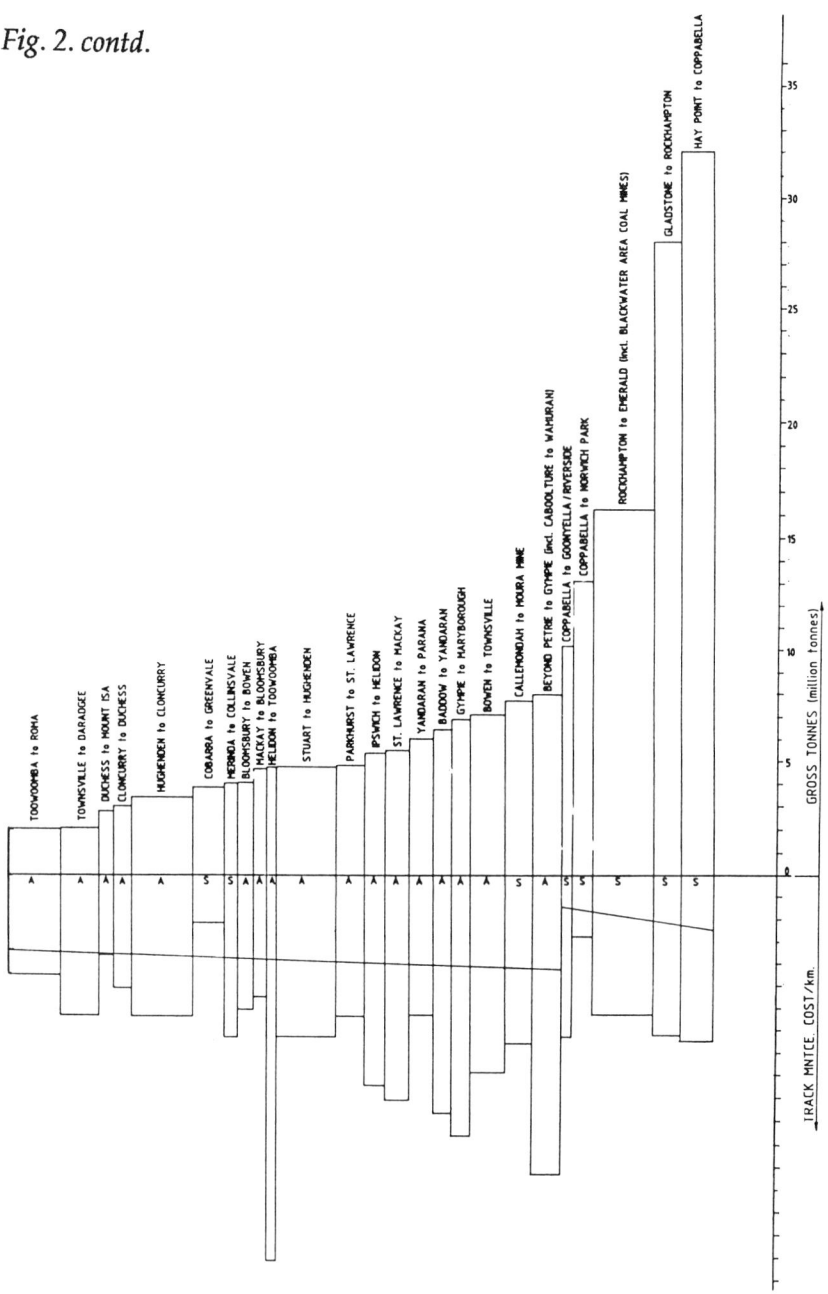

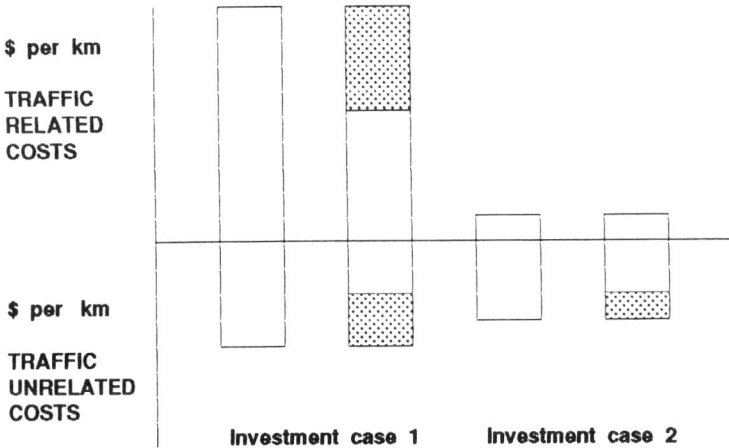

Fig. 3. Infrastructure investments

The graph also shows the trend line between traffic related and traffic unrelated costs. It is apparent that at very low traffic levels, the track maintenance is controlled almost entirely by traffic unrelated costs while the reverse is true where traffic tasks are high.

This type of graph is also useful in identifying sections where maintenance costs are unusually high. Investigation of these locations can quickly identify areas where more appropriate maintenance or investment strategies will produce significant cost benefits.

Where traffic unrelated costs are predominant then investment is more likely to be worthwhile if it is directed towards improving maintenance methods rather than the track structure.

Figure 3 shows how equal investments in the infrastructure may produce varying levels of benefit depending on the traffic related and unrelated costs.

Track condition

QR operates two track recording cars as well as a post analysis system to monitor track condition and to develop the basic maintenance programmes for each route or line section. One of the cars is equipped to monitor the electrified overhead installations.

Freight Task MGT		Category	Expected TCI range			Description
From	To		Lower	Median	Upper	
0.0	0.1	10	48	64	80	Low speed 30 km/h.
0.1	1.0	9	43	58	73	Light B class 30 – 60 km/h.
1.0	3.0	8	39	52	65	Light A class & heavy B class
3.0	8.0	7	34	46	58	Medium A class & suburban
8.0	12.0	6	30	40	50	Heavy A class & medium S class
12.0	50.0	5	25	34	43	Heavy Haul S class

NOTE :

Lower and upper TCI values
are the median value plus
or minus 25 per cent.

Fig. 4. Track condition standards

As may be expected, localised defects are attended to by local maintenance gangs. The track condition over longer lengths forms the basis for
production gang programmes.

To assist in this programming QR uses a post analysis system that can
re-run the recorded information from the track recording cars. In post
analysis it is possible to vary the desired tolerances at will and therefore
identify the production work required to return the track to a predetermined
track condition index.

It can similarly assist in identifying where the most benefit can be gained
with a predetermined work input. An example of this may be to identify
where say 100 hrs of resurfacing should be undertaken to gain the maximum
improvement in track condition on a given track section.

It is apparent that track condition may, and perhaps should, vary with the
traffic task. It is therefore desirable to be able to identify if the level of
resurfacing is at a desirable level for the particular traffic task. QR has
developed a basic standard which defines the desirable track condition for
various gross annual tonnages. (Fig. 4). There will of course be variations
depending on operating speed, degree of curvature etc. The information is
presented in both graphical form and in a tabulation combining track condition, traffic task (in gross tonnes) and track maintenance cost (per km). The

information can give an indication as to whether the maintenance effort should be increased, decreased or maintained at the existing level.

The contribution to track maintenance costs by production gangs can therefore be optimised.

Maintenance equipment

Queensland Rail has progressed from being a predominantly manual maintenance organisation in the mid 1960s to one almost totally dependent on machine maintenance today. In the initial years machines were programmed on a cyclic basis.

In the case of resurfacing, full face resurfacing was programmed each 2 years. With timber sleeper renewals the equipment was programmed over each section each four years. In this case all condemned sleepers and those sleepers not expected to last until the next cycle, were renewed.

The cyclic strategy proved to be quite costly in as much as maintenance managers tended to undertake more work than absolutely necessary. There is no doubt that sleepers were renewed even though they would have been effective in track until the next 4 year cycle. This of course resulted in both unnecessary work being undertaken as well as significantly reducing the average sleeper life.

The additional work also extended the cycle to the extent that track condition deteriorated to an almost unsafe condition between cycles. To add to the problems, mechanical maintenance of the equipment proved to be difficult because of the high turn over of skilled staff, particularly in the more remote areas of the state.

To correct this situation the maintenance strategy was changed to allow for more frequent passes but only undertaking work necessary. There was no nominated cycle. The production gangs were therefore programmed very much on a needs basis. At about the same time QR transferred the mechanical maintenance of the production equipment to contract. The contractor was able to increase financial returns by bonus payments based on machine availability.

This change in strategy has resulted in a marked improvement in track condition so far as resurfacing is concerned, as well as considerably increasing average sleeper life. It has also allowed three resurfacing and three resleepering gangs to be decommissioned with more planned. The cost of track maintenance was reduced by an average of 22% state wide during the period 1983/84 to 1990/91.

The future strategy is to convert to higher production equipment and so reduce both the total equipment inventory and the number of production gangs.

Special equipment
Track laying and re-laying

Because of the amount of new track to be constructed with the expansion of the coal industry as well as the ongoing need to re-lay track, QR acquired a track laying machine. This machine is a Tamper P811S and it was the first machine of this type built to both construct and reconstruct track.

Prior to the purchase of the machine QR undertook most of its track reconstruction using manual gangs supported by a wide range of relatively small plant and equipment. New construction was carried out both by contract, and by QR gangs.

The introduction of concrete sleepers meant that the techniques employed required a serious review.

In considering the acquisition of equipment of this type it is important to realise that its cost-effectiveness is very sensitive to the size of the on going task required of the equipment. If the output cost per kilometre is plotted against the annual task then a series of curves of the general form shown in Fig. 5 is obtained.

The base case, manual work, tends towards a constant cost while mechanised work becomes more cost-effective with greater output.

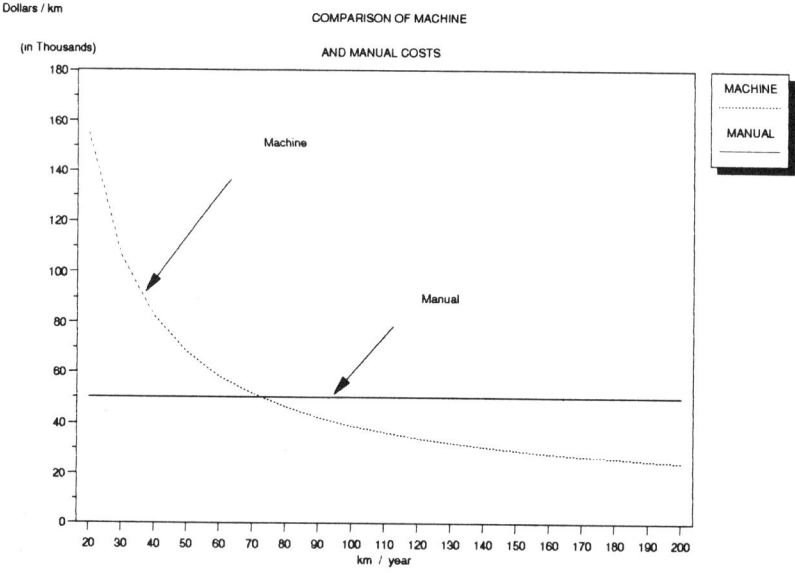

Fig. 5. Re-laying cost

Obviously there are a family of curves depending on the type of mechanised operation being considered. The curve generated will also be system specific. This information can clearly indicate the cost effectiveness of each option at any ongoing output task level. The assessment undertaken by QR showed that the purchase of a track laying machine was the most cost-effective for the required future task (100 km/yr). If the task was less, then contract or semi manual operations were more cost- effective. Since the task was such that the machine would have a full 12 months programme for more than 7 years, QR proceeded with the purchase.

The specification required a machine to both construct and reconstruct track. It also required an axle load of no more than 12 t. Timber, steel and concrete sleepers had to be handled through the machine with minimal conversion time. In the period 1984-1991, 71 km of new track has been constructed and 613 km of relaying has been completed. The operation remains very cost-effective on all classes of track.

Rail rectification

QR operates a rail planing machine and a 32 stone rail grinder. This equipment is used predominantly on the heavy coal lines to provide and maintain asymmetric rail profiles to reduce both rail and wheel wear. Both units are used to restore standard profiles on freight lines as well as for the removal of rail corrugation and other surface defects.

The asymmetric grinding has resulted in an extension of rail life of over 100% and wheel life has been extended by some 10 times.

The benefits to rail management provided by equipment of this type are well documented, particularly where gross annual tonnages and high axle load traffic is involved.

Ballast cleaner/undercutter

Until recently, QR had not undertaken ballast cleaning. The conditions prevailing on most lines were such that ballast replacement was required rather than cleaning. Replacement was carried out by removing the old ballast material using an undertrack sled and simply replacing the ballast using ballast hopper wagons.

In recent years however, contamination of the ballast by coal particles on the heavy haul lines demanded that past policy be reviewed. Analysis of the ballast showed that in extensive sections of track, the ballast was in a condition such that reclaiming was simply not of benefit.

The cleaning operations were required to be undertaken in tropical locations suffering heavy and sustained wet seasons. Ballast cleaning in these conditions is not particularly effective because of screening problems as well as the retention of excessive amounts of fine material on the ballast.

QR finally proceeded with the purchase of a combination ballast cleaner and undercutter. This machine allows for the high speed removal of all ballast as well as having the capacity to clean ballast.Each section of track is assessed to determine the most cost-effective method of treatment prior to the cleaning/undercutting being programmed.

At present the track-laying machine is operating on the heavy haul lines and the ballast cleaner/undercutter is being used as part of the total track reconditioning operation.

Investigations are being undertaken to determine the cost effectiveness of shoulder ballast cleaning as it may apply to QR. It is thought that this operation is likely to produce the results required on most of the system's general freight lines.

Natural hazards

Queensland's climate varies from tropical to warm temperature. It is subject to intense monsoonal and cyclonic rain through the summer months. During this period, flooding, and resultant track damage is quite common. In range areas land slips can affect the track.

Many of the QR's bridges are at a low level and are frequently flooded. The track traverses major flood plains, some more than 100 km wide. Widespread flooding and extensive damage can be sustained in these locations. Repairs to flood damaged track is costly, time consuming and labour intensive. Loss of revenue can be substantial.

On-going capital investment in high level bridges has significantly reduced the interruption to services as well as reducing the level of flood damage over the major routes. Bridging over many of the flood plains however is simply not practical. Lifting the level of the track could be of some benefit, however the increased afflux would cause significant problems to communities and properties upstream of the embankment.

In 1984 Queensland Rail sponsored a research project at the University of Queensland directed towards finding a cost-effective means of preventing or limiting damage due to flooding. The results of the study indicated that the track could be flood proofed to the extent that service recovery could be effected almost immediately after the flood water had cleared from the track. Fig. 6 shows how this was achieved.Trail installations proved that the results predicted were achievable. To date over 100 km of track have been flood-proofed in an ongoing programme. It will be many years before completion of the programme, however the benefits in both revenue earnings and reduced maintenance expenditure are already at a level that will encourage increased investment.

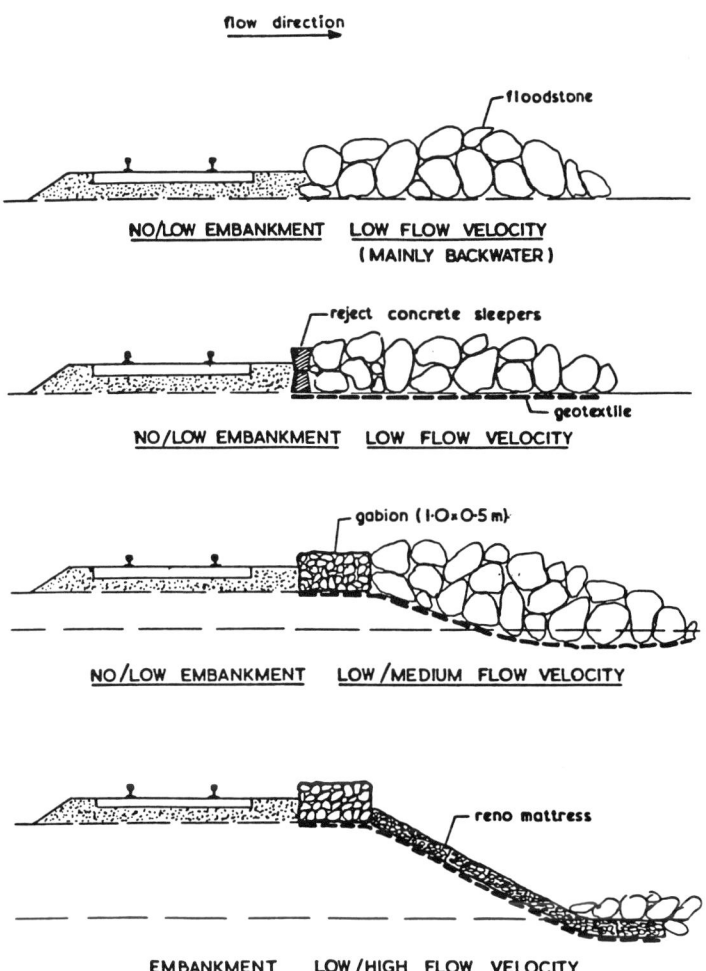

Fig. 6. Flood-proofing systems

Organisational structure

As suggested earlier, the structure of the maintenance organisation can have a major impact on the track maintenance costs as well as on the effectiveness of the maintenance effort. Until recently, restructuring of QR's maintenance organisation was an evolutionary process. Even with the introduction of mechanised maintenance equipment there was always a reluctance to move too far away from the comfort of having many local gangs stationed fairly close together. The mechanised maintenance was virtually superimposed upon the old structure. Cost savings were generally achieved by staffing the local gangs at lower levels together with some amalgamation of gangs. As previously suggested track maintenance costs have been reduced by some 22% statewide in the period 1983/84 to 1990/91.

With the total restructure of Queensland Rail from the top down, the time was right to reconsider the maintenance organisation's role in the new commercial structure.

Extensive comparisons were made of QRs track maintenance input and output using data available through the Railways of Australia Organisation and data available on some international railways. It was no real surprise that QR did not show up very well in many of the performance indicators considered, particularly on the older freight lines.

The available data suggested quite clearly that resource input in the way of labour, equipment and material was far too great for the outputs being obtained.

It was also evident that maintenance work was being programmed to suit the resources available rather than the resources being programmed to meet the required output level for an efficient traffic operation.

It was apparent that further modification of the existing structure would not produce significant efficiencies. Such modifications are rarely sufficient to break out of the maintenance manager's comfort zone.

It is also significant that at about the same time extensive restructuring of Australian industry was being undertaken along guidelines developed by the Federal Industrial Commission. The guidelines set down the principles for wage increases tied to structural efficiency and award restructuring. As a part of this exercise QR's civil maintenance organisation, through consultative groups consisting of management, union and employee representatives, had identified where significant productivity gains could be made at the workface. Desirable structural changes to the organisation were also identified. As a result of this work, a complete maintenance district was converted to operate under the proposed new structure for a trial period of 6 months. The results were impressive with productivity gains in some functions in excess of 100% being recorded.

The information gained from the trial was able to be used in developing the final structure for QR's new maintenance organisation.

The new structure requires far fewer work units (gangs), however the work units are generally larger and much more mobile than was previously the case. Each work unit has in-built multi-skilling as far as practical therefore largely eliminating the need for specialist work units. Each work unit can be subdivided to undertake specific tasks on a daily basis. This allows both the skills and the resources to be more closely matched to the task.

Major production units, while being fewer in number, are also more mobile. They are programmed on a long-term basis, but only undertake work on a needs basis.

The new structure significantly reduces the management and supervisory task. Many of the decisions necessary can be made at the work unit level or at district level. This will allow the management and supervisory staff to be reduced by 50%.

The new structure also lends itself to ongoing further major organisational changes. The most dramatic changes will occur in the Freight Infrastructure Group. This Group maintains approximately 6800 route kilometres of predominantly single track, or over two-thirds of the total state system. Fig.7 shows the effect of the restructure on the field organisation.

Figs 8 and 9 show the percentage changes to the infrastructure indicators of total staff and total cost, expected during the coming five year period.

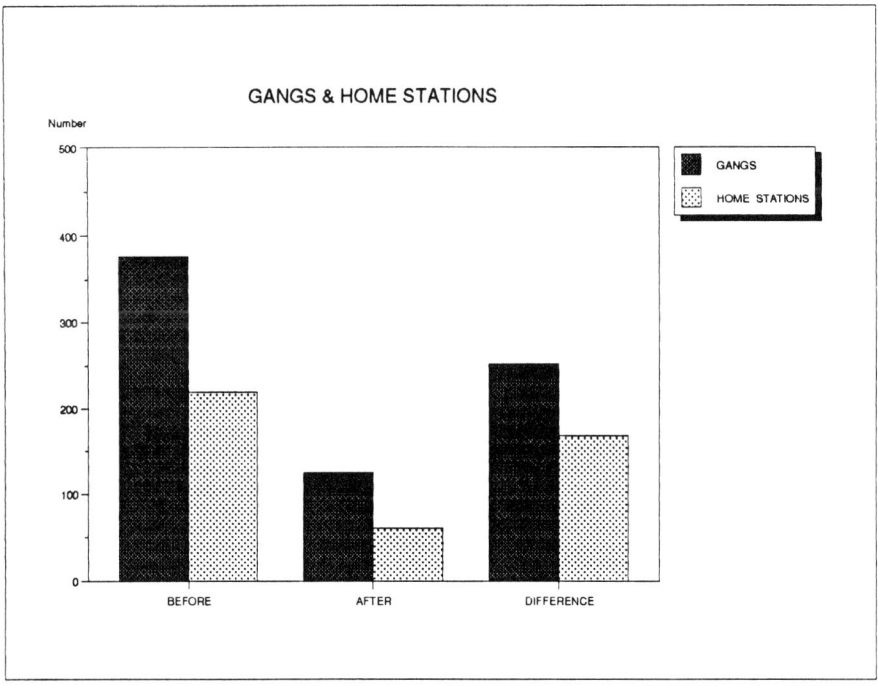

Fig. 7. Gangs and home stations

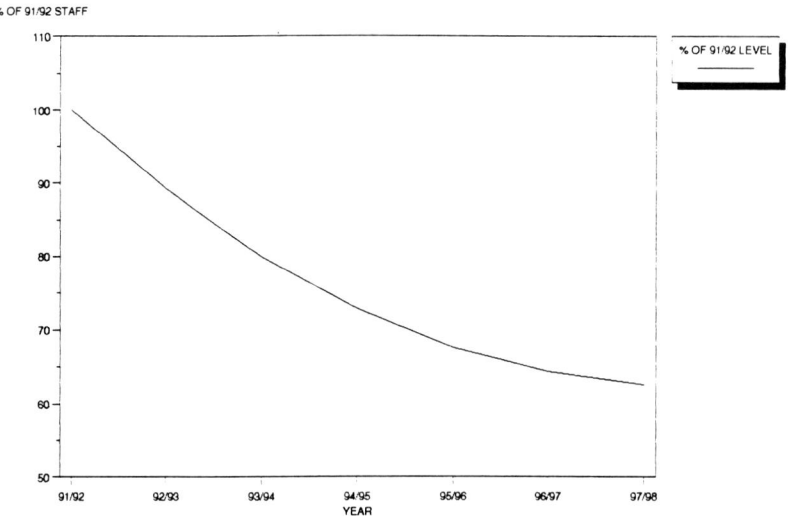

Fig. 8. Total staff changes

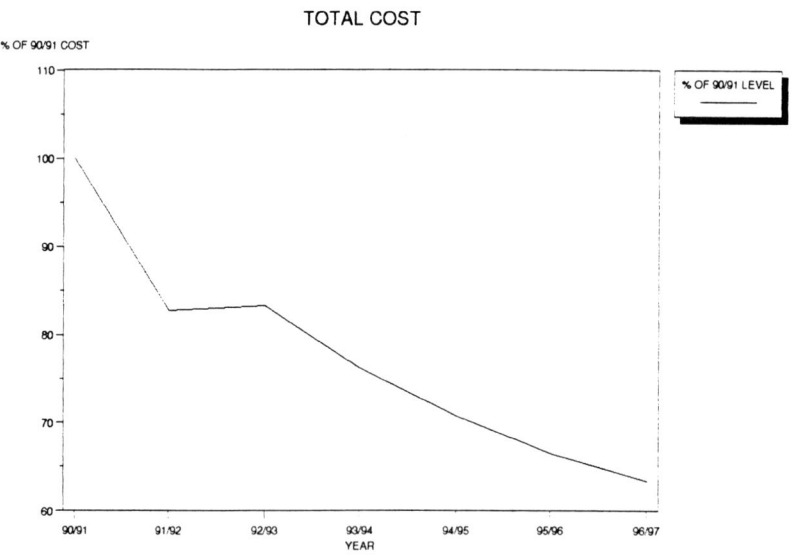

Fig. 9. Total costs

The changes are being implemented as rapidly as possible, however it is expected that it will take approximately 5 years before the final staffing levels are reached. Both the Queensland Government and QR have agreed that staff will not be dismissed nor will they be required to relocate involuntarily.

A range of incentives have been made available to staff to encourage them to decide on their futures. These incentives include attractive relocation assistance, retraining assistance and an attractive voluntary early retirement package.

The changes being introduced will have a dramatic and positive contribution to the bottom line performance of Queensland Rail without degrading the operational efficiency of the track structure.

Conclusions

The major changes occurring within Queensland Rail have caused the past maintenance strategies and structures to be questioned. There is no doubt that the changes being implemented will be of huge benefit to QR in the future. It has become quite apparent during this exercise that the major benefits will come from the changes to both management thinking and management attitudes. It is essential that managers review and continue to review their organisations. They must put in place organisations and systems that will allow higher productivity to be achieved. They must also be sure of the value of investment and ensure that the benefits are achieved.

Cost-effective maintenance is first and foremost a cost-effective maintenance manager.

Discussion on Papers 13 –15

F. I. MAU, Vice-President Operations, BHP Rail Products (Canada) Ltd
Steel sleepers of appropriate design can be very effectively utilized on a 1 in 4 replacement of timber sleepers basis. The life of the timber sleepers is extended and gauge of track is assured. This method of cost-effective maintenance has been used for some time in Australia and is becoming popular in North America.

M. FULKER, London Underground Ltd
In 1987, I was seconded to LTI and RITES to visit Calcutta Metro as a permanent way consultant to review construction, design maintenance and training associated with recent (at that time) problems during construction.
I would be interested to know whether the project has been completed and is fully operational, and what lessons have Indian Railways learned from constructing its first metro.

H. M. AHMED, former General Manager, Sudan Railways
The slotting of steel sleepers between timber sleepers is a very interesting and beneficial practice which can give more life span to the timber sleepers in the track.
But it is necessary to be very careful about the design especially in hot weather as the sleepers have two different sets of characteristics. A design that neglected such considerations could be detrimental to the track and may lead to track irregularities and hence accidents.
Could Mr Bell elaborate on the use of this practice in Australia ?

F. R. BELL, Author
Queensland Rail has been installing steel sleepers in a one in four pattern among timber sleepers on lines with less than 8 mgt.
This patterned replacement has been trialled in several Australian rail systems following extensive research undertaken by Melbourne Research Laboratories, a division of BHP Australia. The research and subsequent trials have shown that the track structure is strengthened and shows greater stability with this one in three or one in four pattern. It is essential, however, that the end design of the sleeper is such that it will resist lateral movement.

DISCUSSION

Modifications to the sleeper design over the past few years by BHP has resulted in a sleeper that is compatible with timber sleepers and provides good lateral stability.

On the more heavily used lines, Queensland Rail is intending to continue with steel sleeper insertions to the stage where the one in four pattern will eventually be converted to fully steel sleepered track.

In reply to H. M. Ahmed the design of the steel sleeper, particularly the end shape, is critical to the success of patterned steel sleeper insertions.

Trials in several Australian Railways have shown that steel is compatible with timber and can increase the strength and stability of the track.

Most of Queensland Rail's system is in tropical and sub tropical regions and extends in to the arid, semi desert, western areas of the state. Temperature ranges are considerable.

Where one in four steel insertions have been in use for several years on Queensland Rail tracks, there has been no incidence of misalignments due to temperature.